D0243550

achieve
grade
A

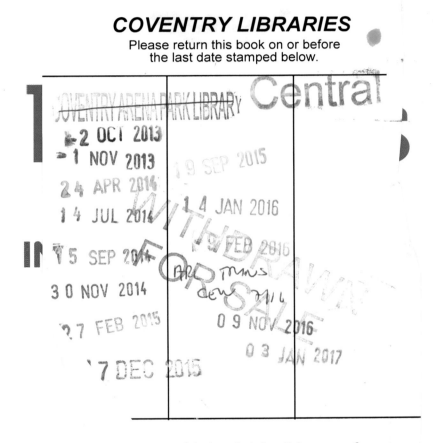

GALORE PARK

www.galorepark.co.uk

Published by Galore Park Publishing Ltd
19/21 Sayers Lane, Tenterden, Kent TN30 6BW
www.galorepark.co.uk

Design and typesetting River Design
Technical illustrations by Ian Moores

Printed by Charlesworth Press, Wakefield

ISBN 978 1 905735 563

First published 2012

Details of other Galore Park publications are available at www.galorepark.co.uk

The publishers are grateful for permission to use the photographs as follows:

P23 sciencephotos/Alamy; P37 NASA 1997, by kind permission of The Space Telescope Science Institute (STScI); P43 whiteboxmedia limited/Alamy; P44 Paul Reid/shutterstock; P45 Flake/Alamy; P49 Stockbyte/ Photos.com; P55 Photoroller/shutterstock; P55 Boris Sosnovyy/shutterstock; P70 sciencephotos/Alamy; P70 sciencephotos/Alamy; P101 zts/Photos.com; P105 sciencephotos/Alamy; P135 Gjermund Alsos/shutterstock; P152 charistoone-stock/Alamy; P159 Andrew Lambert Photography/Science Photo Library; P160 Gerrit/ shutterstock; P178 Martyn F. Chillmaid/Science Photo Library; P183 Health Protection Agency/Science Photo Library; P198 Solodov Alexey/shutterstock.

Contents

Introduction

Physics is the basis for much of the technology we use in everyday life, and also for many more specialised technologies ranging from medical imaging equipment to space probes. A good grounding in physics is a useful background not only to careers in science and engineering, but also in many areas where a degree of technical understanding is needed. Physics, chemistry and biology are often seen as separate subjects, but all three are important in different ways and they have many connections with each other.

This revision guide presents the facts as clearly and as logically as possible to help you to gain both knowledge and understanding. You will learn best if you use this book interactively: make your own notes on paper as you go along, or highlight key terms or concepts that you struggle with. You may like to make spider diagrams to help you to link ideas together or summarise chapters. Find out which study methods work best for you – and good luck with your studies and your exams!

How is this book organised?

This book follows the structure laid out in the Edexcel International GCSE Physics specification (4PH0) and, where necessary and appropriate, includes extra material to fulfil the Cambridge International Examinations (CIE) International GCSE Physics specification (0625).

Material that is only applicable to CIE students is indicated by this symbol.

Each section begins with a list of what you are expected to know for that particular topic, followed by the material that you need to learn. This has been laid out in a revision-friendly way, using bulleted lists and tables to aid visual learning. Worked examples of calculations are also included. You should work through this material in the way that suits you best. You may choose to read it aloud to yourself or a friend, or to write it out in longhand. The material in this book gives you the essential points of the topic, but remember that to revise effectively you will need to rework it, either mentally or on paper, into concise factual notes that you will be able to remember under exam conditions.

Throughout the book you will find tip boxes that will help you to achieve the A* grade. Some of these apply to all topics; others concentrate on a particularly tricky piece of theory, or something that often causes candidates to trip up.

Towards the end of each section is a review box that acts as a checklist. Once you have worked through the section, check that you can do everything listed in the box. If not, use the page references to refer back to the text and learn that part again. There are also some review questions after the review box to ensure you have understood everything and to help highlight any areas that may need revisiting.

Each section concludes with a set of practice questions. These are written in the style you will encounter in your exam paper and are designed to encourage analysis of ideas which are integral to achieving an A* grade. The answers are at the back of the book. Practice makes perfect, so once you have completed the questions in this book, get hold of some actual past papers.

Key words have been printed in **bold**. You should be sure you understand their meaning.

How will I be assessed?

Edexcel

Edexcel candidates take two exam papers:

- Paper 1 lasts 2 hours and is worth two-thirds of the overall mark.
- Paper 2 lasts 1 hour and is worrth one-third of the overall mark.

Edexcel candidates are not assessed through coursework.

Successful candidates must meet all three assessment objectives:

AO1   Knowledge and understanding
AO2   Application of knowledge and understanding, analysis and evaluation
AO3   Investigative skills (from June 2013 AO3 will change to: Experimental skills, analysis and evaluation of data and methods)

CIE

CIE candidates take **two** papers from the following:

- Paper 1, a multiple choice paper lasting 45 minutes and worth 30% of the overall mark
  and either
- Paper 2, a core curriculum paper lasting 1 hour 15 minutes and worth 50% of the overall mark
  or
- Paper 3, an extended curriculum paper lasting 1 hour 15 minutes and worth 50% of the overall mark.

Plus **one** of the following, all of which are worth 20% of the overall mark:

- Paper 4, coursework conducted at your school
- Paper 5, a practical test lasting 1 hour 15 minutes, conducted at your school
- Paper 6, a written paper on practical theory lasting 1 hour.

Successful candidates must meet all three assessment objectives:

A Knowledge with understanding
B Handling information and problem solving
C Experimental skills and investigations

Some help with revision

The most common error is to think you have revised well just because you have spent a long time on it. The following advice should help you to revise effectively:

- Never work for more than 30 minutes at a stretch. Take a break.
- Don't revise all day. Divide the day into thirds (you will have to get up at a sensible time) and work two-thirds of the day at most.
- Always start your revision where you finished your last session and briefly review what you covered before. You absorb facts better if you always meet them in more or less the same order.
- Don't revise what you already know. That's like practising a complete piano piece when there is only a short part of it that is causing you problems.
- Annotate your revision notes and make use of coloured highlighting to indicate areas of particular difficulty.
- Do some of your revision with other students. The knowledge that others find parts of the syllabus tricky can be comforting and other students may well have found an effective way to cope, which they can pass on to you.
- Finally, remember that International GCSE physics has no hidden pitfalls. If you learn the facts and understand the principles, you will secure the high grade you deserve.

Also available in the A* range

English A* Study Guide for GCSE and IGCSE by Susan Elkin, ISBN 978 1 905735 426

Chemistry A* Study Guide for International GCSE by Frank Benfield, ISBN 978 1 905735 440

Biology A* Study Guide for International GCSE by Pamela Maitland, ISBN 978 1 905735 457

All available from Galore Park, www.galorepark.co.uk

Section One

Forces and motion

A Movement and position

You will be expected to:

★ understand what distance–time graphs show and use information from them
★ recall and use the formula relating speed, distance and time to each other, including rearranging it if necessary
★ understand the difference between vector and scalar quantities
★ recall and use the formula relating acceleration, velocity and time to each other, including rearranging it if necessary
★ understand what velocity–time graphs show
★ use information from a velocity–time graph to work out acceleration
★ work out the distance travelled from the area under a velocity–time graph.

TIP If the specification says *recall*, it means you need to remember that formula. It will not be given to you in the exam.

Units

You will use the following quantities and units in this section. You should already know most of these from your previous studies.

Quantity	Symbol	Unit
acceleration	a	metres per second^2 (m/s^2)
distance	d	metres (m)
initial velocity	u	metres per second (m/s)
speed or velocity	s or v or u	metres per second (m/s)
time	t	seconds (s)

Distance–time graphs

Distance–time graphs show how the distance of an object from a starting point changes with time.

• A horizontal line shows that the object is not moving.
• A sloping line shows the object moving at a constant speed.
• Steeper lines show higher speeds.

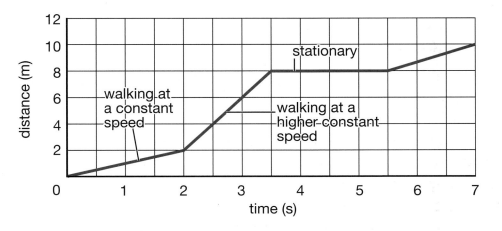

Fig. 1a.01: Distance–time graph for a person walking

Calculating speed

The **speed** of an object is the distance it moves in a certain time.

$$speed = \frac{distance}{time} \qquad s = \frac{d}{t}$$

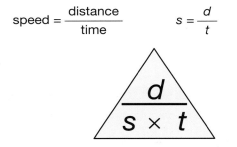

**Fig. 1a.02: Formula triangle for speed, distance and time;
Appendix 1 shows you how to use formula triangles**

The standard units for speed in physics are metres per second (m/s). Units for speed used in everyday life include miles per hour (mph) and kilometres per hour (kph or km/h). You may need to use these units in the exam. Just remember that the units for speed depend on the units for distance and time, so if you are given a distance in kilometres and a time in hours, the speed you calculate will be in km/h.

Worked example

How fast is the person in the graph in Fig. 1a.01 (page 3) walking in the first 2 seconds?

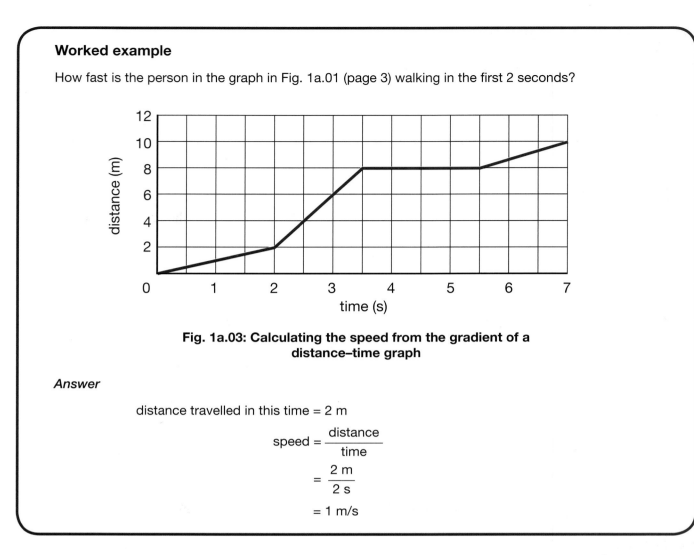

Fig. 1a.03: Calculating the speed from the gradient of a distance–time graph

Answer

distance travelled in this time = 2 m

$$\text{speed} = \frac{\text{distance}}{\text{time}}$$

$$= \frac{2\ m}{2\ s}$$

$$= 1\ m/s$$

Average speed is the total distance for a journey divided by the total time. In Fig. 1a.01, the average speed for the whole journey is worked out using the total time (7 seconds) and the total distance (10 metres).

The speed formula can be rearranged to allow you to calculate distance or time.

$$\text{distance} = \text{speed} \times \text{time}$$

$$\text{time} = \frac{\text{distance}}{\text{speed}}$$

TIP

You need to remember this formula as it will not be given to you in the exam. You also need to be able to rearrange it, if you are asked to calculate a distance or a time. It helps if you understand how to rearrange formulae (see Appendix 1), but if you struggle with this you can either memorise the three different versions, or remember the formula triangle shown in Fig. 1a.02.

Measuring speed

The speed of an athlete in a race can be measured by:

- timing how long it takes to complete the distance of the race
- calculating the average speed using the formula on page 4.

The speed of smaller objects can be measured in the laboratory using a light gate and a datalogger (Fig. 1a.04) or using a ticker timer (Fig. 1a.05).

Light gate and datalogger

The card on the trolley cuts the light beam. The datalogger measures the time it takes the card to go through the gate. The speed is worked out from the length of the card and the time.

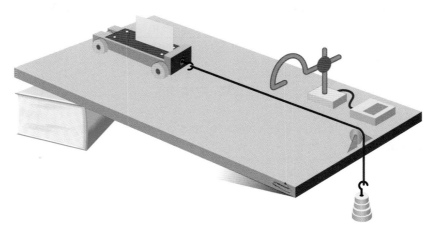

Fig. 1a.04: Measuring speed using a light gate and datalogger

Ticker timer

The ticker timer makes a dot on the paper tape 50 times each second. You can work out the speed by counting the number of dots in a certain length of tape. The number of dots tells you how long it took that length of tape to be pulled through the machine.

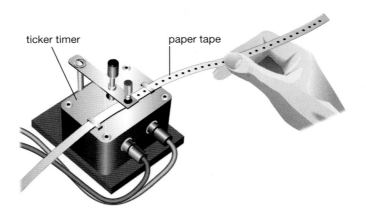

Fig. 1a.05: Measuring speed using a ticker timer

Scalar and vector quantities

Speed is a **scalar** quantity. It has a magnitude but not a direction. **Velocity** is a **vector** quantity. A car travelling around a bend can have a constant speed but not a constant velocity, as its direction is changing all the time. Scalar and vector quantities are covered in more detail in Section 1B, page 14.

Calculating acceleration

Acceleration is how fast the velocity of an object is changing. The units for acceleration are metres per second per second, or m/s^2.

A sports car that can go from 0 to 100 km/h in 2.5 seconds has a much greater acceleration than a family car that takes 15 seconds to get to 100 km/h.

Acceleration is calculated using this formula:

$$acceleration = \frac{change\ in\ velocity}{time\ taken}$$

$$a = \frac{(v - u)}{t}$$

TIP
You need to remember this formula as it will not be given to you in the exam.

where v = final velocity and u = initial velocity

An object that is slowing down has a *negative* acceleration. This is sometimes called a **deceleration**.

Worked example

A car accelerates from 20 m/s to 25 m/s in 2.5 seconds. What is its acceleration?

Answer

$$a = \frac{(25\ m/s - 20\ m/s)}{2.5\ s}$$

$$= \frac{5\ m/s}{2.5\ s}$$

$$= 2\ m/s^2$$

TIP
This formula is not quite as easy to rearrange as the formula for speed. If you are asked to calculate a time or a change in velocity, it is probably best to remember the version in words shown in Fig. 1a.06.

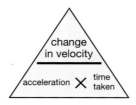

Fig. 1a.06: Formula triangle for the change in velocity

Worked example

A car accelerates at 2 m/s² for 5 seconds, reaching a speed of 15 m/s. What was its initial velocity?

Answer

Rearranging the formula for acceleration:

$$\text{change in velocity} = \text{acceleration} \times \text{time taken}$$
$$= 2 \text{ m/s}^2 \times 5 \text{ s}$$
$$= 10 \text{ m/s}$$

$$\text{initial velocity} = \text{final velocity} - \text{change in velocity}$$
$$= 15 \text{ m/s} - 10 \text{ m/s}$$
$$= 5 \text{ m/s}$$

Velocity–time graphs

Velocity–time (v–t) graphs show how the velocity of an object changes with time.

- A horizontal line shows that the object is moving at a constant speed.
- A line sloping upwards shows that the object is accelerating.
- The steeper the line, the greater the acceleration.
- A line sloping downwards shows that the object is decelerating (it is slowing down).

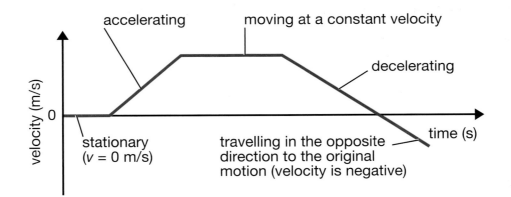

Fig. 1a.07: Velocity–time graph

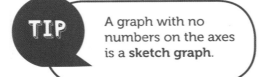

TIP   A graph with no numbers on the axes is a **sketch graph**.

Make sure you are clear about the differences between distance–time and velocity–time graphs. In an exam, read the question carefully to check which kind of graph you are looking at.

Line	On a distance–time graph means ...	On a velocity–time graph means ...
horizontal line along the time axis	stationary – has not moved from the starting point	stationary
horizontal line	stationary	moving at a constant velocity
sloping upwards	moving at a constant velocity – the gradient of the line gives the velocity	accelerating – the gradient of the line gives the acceleration

Calculating acceleration from a *v–t* graph

Acceleration is calculated using a change in velocity and a time. This information can all be obtained from a velocity–time graph.

Worked example

A car accelerates when it reaches the motorway, as shown in Fig. 1a.08. What is its acceleration?

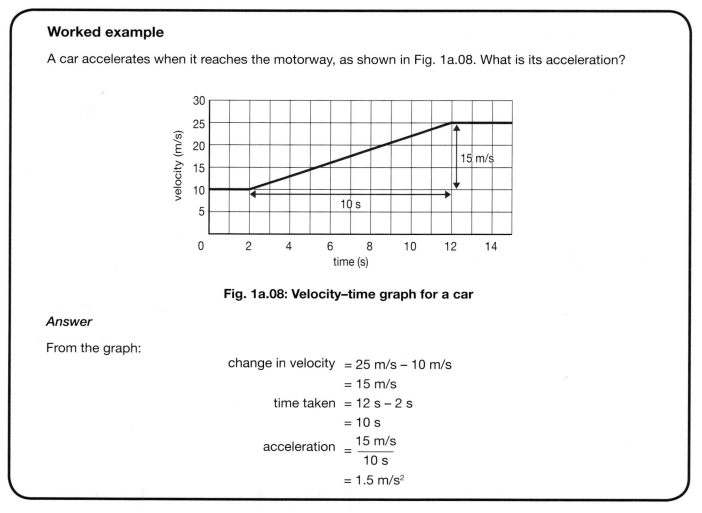

Fig. 1a.08: Velocity–time graph for a car

Answer

From the graph:

$$\text{change in velocity} = 25 \text{ m/s} - 10 \text{ m/s}$$
$$= 15 \text{ m/s}$$
$$\text{time taken} = 12 \text{ s} - 2 \text{ s}$$
$$= 10 \text{ s}$$
$$\text{acceleration} = \frac{15 \text{ m/s}}{10 \text{ s}}$$
$$= 1.5 \text{ m/s}^2$$

The vertical difference between two points on a graph divided by the horizontal difference is the **gradient** of the line. The gradient of the line on a velocity–time graph gives the acceleration.

Physics A* Study Guide

Measuring acceleration

A ticker timer can be used to measure acceleration. Fig. 1a.05 (on page 5) shows how a ticker timer works, and Fig. 1a.09 (below) shows how the ticker tape is used. Although you are sticking *lengths* of paper on the graph, each length represents a distance covered *in a certain time*, so it represents a velocity, not a distance.

(a)

(b)

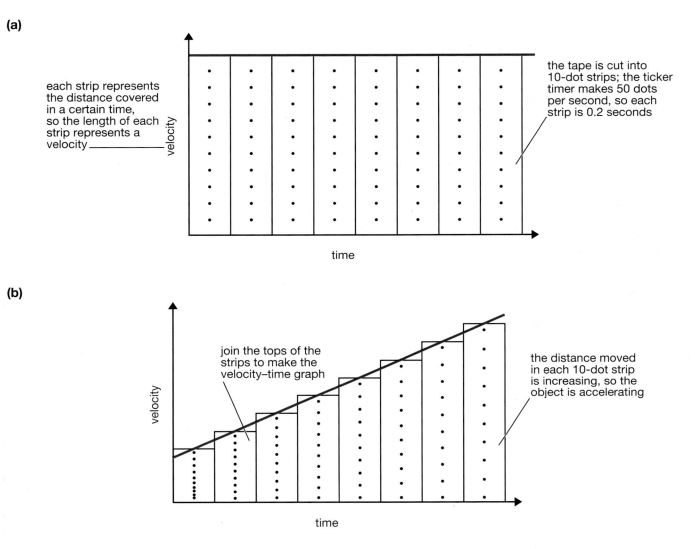

Fig. 1a.09: Velocity–time graphs made out of strips of ticker tape:
(a) This graph represents an object moving at a constant speed; it moves the same distance
in each time interval
(b) This graph represents an object that is accelerating; the length of each strip represents the velocity,
and the velocity is increasing with time

Calculating distance travelled from a *v–t* graph

Fig. 1a.09 (on page 9) shows two velocity–time graphs made using ticker tape. The total length of the ticker tape stuck on each graph represents the total distance moved by the object. This is also the area between the line on the graph and the time axis.

You can work out the distance moved by any object represented on a velocity–time graph by using a combination of rectangles and triangles.

> area of rectangle = length × width
> area of triangle = $\frac{1}{2}$ × base × height

Worked example

How far does the car shown in Fig. 1a.08 (on page 8) move in the first 12 seconds shown on the graph?

Fig. 1a.10: Calculating the distance moved from the area under a velocity–time graph

Answer

Distance moved = area under graph

= area A + area B + area C (see Fig. 1a.10)

= (2 s × 10 m/s) + ($\frac{1}{2}$ × 10 s × 15 m/s) + (10 s × 10 m/s)

= 20 m + 75 m + 100 m

= 195 m

TIP

Remember that for many questions which involve reading values from graphs, it is the difference between two values you need, not just the numbers on the axis.

For instance, the base of triangle B in Fig. 1a.10 is 10 seconds long, not 12 seconds.

You should now be able to:

★ sketch distance–time graphs to show (a) an object moving at a steady speed, (b) a stationary object (see page 3)

★ recall and use the formula relating speed, distance and time, rearranging it if necessary (see page 3)

★ recall and use the formula relating acceleration, velocity and time, rearranging it if necessary (see page 6)

★ sketch velocity–time graphs to show objects that are (a) decelerating, (b) accelerating, (c) stationary, (d) travelling at a constant speed (see page 7)

★ describe in words how you would use a velocity–time graph to calculate (a) acceleration, (b) distance travelled (see page 8).

Review questions

1. In Fig. 1a.01 (on page 3), what is the speed of the walker in the final section of the graph?

2. A runner completed a 200 m race at an average speed of 6 m/s. For how long was he running?

3. (a) A cyclist moving at 2 m/s changes speed to 7 m/s over a time of 2 seconds. What is her acceleration?

   (b) A cyclist accelerates from a speed of 0.5 m/s at 2 m/s². What is her speed after 3 seconds?

4. (a) Describe the movement of the object during the first 25 seconds of the journey shown in this velocity–time graph.

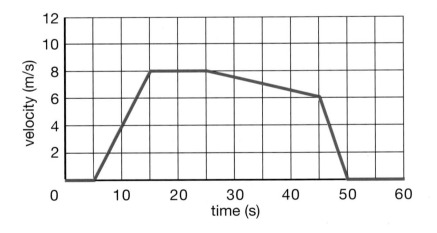

(b) What is the acceleration between 25 and 45 seconds?

(c) How far does the object travel during the period when it is accelerating to 8 m/s?

Practice questions

1. The graph shows the distance a rambler travelled during a hill-walk.

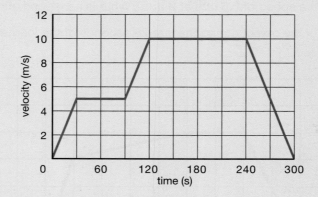

(a) How long after starting did the walker stop for lunch? Explain your answer. **(2)**

(b) Part of the walk involved going up a steep hill. Suggest where in the journey this was, and explain your answer. **(2)**

(c) How fast was the walker moving for the first 1.5 hours of her journey? Show all your working. **(4)**

2. A car is used to drive through town. The velocity–time graph shows its motion.

(a) How do you know the car is accelerating between 0 and 30 seconds? **(1)**

(b) Calculate the acceleration of the car between 240 and 300 seconds. **(4)**

(c) How can you work out the total distance the car travels during the journey? **(1)**

(d) Calculate how far the car travels in the first minute of its journey. **(5)**

B Forces, movement, shape and momentum

You will be expected to:

* identify different types of force, including friction and weight
* explain the difference between vector and scalar quantities
* add forces that act along a line
* **(CAM)** find the resultant of two forces that do not act along the same line
* recall and use the formula relating force, mass and acceleration, including rearranging it if necessary
* recall and use the formula relating weight, mass and the acceleration due to gravity, including rearranging it if necessary
* describe the forces acting on falling objects
* explain why falling objects reach a terminal velocity
* describe the factors affecting vehicle stopping distances
* recall and use the formula relating momentum, mass and velocity, including rearranging it if necessary
* use ideas about momentum to explain safety features
* use conservation of momentum to calculate the mass, momentum or velocity of objects
* recall and use the formula relating force, change in momentum and time taken, including rearranging it if necessary
* understand Newton's third law (action and reaction forces)
* recall and use the formula for calculating moments
* use the principle of moments to calculate forces on objects in equilibrium
* explain the centre of gravity of a body
* **(CAM)** explain how to find the centre of mass of an irregular shape
* describe how the position of the centre of mass affects the stability of an object
* explain what elastic means
* describe how the extension of springs, wires and rubber bands changes when the force on them changes
* recall Hooke's law.

Units

You will use the following quantities and units in this section. You should already know most of these from your previous studies.

Quantity	Symbol	Unit
acceleration	a	metres/second^2 (m/s^2)
extension	x	metres (m)
force	F	newtons (N)
mass	m	kilograms (kg)
moment	M	newton metres (N m)
momentum	p	kilogram metres/second (kg m/s)
velocity	v	metres/second (m/s)
weight	W	newtons (N)

Forces

A **force** is a push or pull exerted on one body by another.

Contact forces

Some forces act when two bodies are in contact with each other, such as:

- the pull of an extended spring
- **friction** acting to slow down moving objects.

Friction can occur between two solid objects, or between a solid object and a fluid. Air resistance and water resistance are both forms of friction. The direction of the friction force opposes the motion of the objects or fluids.

Forces at a distance

Some forces act at a distance. These include:

- gravitational force between bodies (weight)
- magnetism
- electrostatic force between electrical charges.

Forces can change the velocity or shape of an object. Changing the velocity includes:

- changing the object's speed
- starting or stopping its movement
- changing the direction in which it moves.

Vector and scalar quantities

Scalar quantities have a magnitude. Scalar quantities include speed, mass, length, volume, density, etc.

Vector quantities have a magnitude *and* a direction. A change in a vector quantity can involve a change in the magnitude, a change in the direction, or a change in both. Vector quantities include forces, velocity, acceleration and momentum.

Adding vector quantities

When you are asked to add vector quantities, you need to take account of the direction as well as the magnitude.

If two forces are acting in the same direction:

- the magnitude of the **resultant** force is the sum of the magnitudes of the two forces
- the resultant acts in the same direction.

If the forces are in opposite directions:

- the magnitude of the resultant is found by subtracting one force from the other
- in this case it helps to think of one direction being the 'positive' direction.

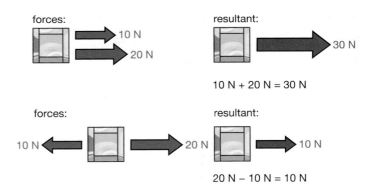

10 N + 20 N = 30 N

20 N − 10 N = 10 N

Fig. 1b.01: Finding resultants of forces acting along a line by adding or subtracting their magnitudes

CAM

Adding vectors using scale drawings

You can find the resultant of two forces at an angle to each other using a scale drawing, as shown in Fig. 1b.02.

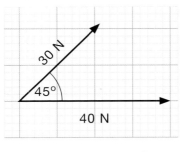

1 cm represents 10 N

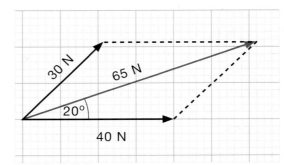

Fig. 1b.02: Scale drawing of vectors

In the diagram, a 30 N force is acting at an angle of 45° to a 40 N force. You find the resultant of these two forces by drawing them to scale and then drawing lines parallel to the forces to complete a parallelogram.

The resultant force is the diagonal of the parallelogram. In the diagram, the resultant force is 65 N at an angle of 20°.

TIP

When using scale drawings to work out resultant forces:

- pick a sensible scale for your drawing – you will get a more accurate result from a bigger drawing
- draw the lengths and angles of the lines as accurately as possible
- use a pencil, not a pen, as it will be easier to correct mistakes
- you may find it easier to do your scale drawing on graph paper.

Force, mass and acceleration

A force can change the velocity of an object. A change in velocity is an acceleration.

The acceleration produced by a particular force depends on the mass of the object: the greater the mass, the smaller the acceleration.

Force, mass and acceleration are related by this formula:

$$\text{force} = \text{mass} \times \text{acceleration}$$
$$F = m \times a$$

The formula can be rearranged to allow you to calculate mass or acceleration:

$$\text{mass} = \frac{\text{force}}{\text{acceleration}}$$

$$\text{acceleration} = \frac{\text{force}}{\text{mass}}$$

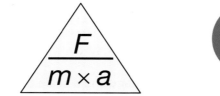

TIP — You need to remember this formula as it will not be given to you in the exam.

Fig. 1b.03: Formula triangle for force, mass and acceleration

---

Worked example

A wheeled suitcase with a mass of 10 kg is pushed with a force of 20 N. What is its acceleration?

Answer

$$\text{acceleration} = \frac{\text{force}}{\text{mass}}$$

$$= \frac{20\text{ N}}{10\text{ kg}}$$

$$= 2\text{ m/s}^2$$

Both force and acceleration are vector quantities.

- If a moving object has a force acting on it in the direction opposite to its movement, the acceleration will act to slow it down.
- If you have used the direction of movement as the positive direction, a force acting in the opposite direction will have a negative sign and will produce a negative acceleration (a deceleration).

Fig. 1b.04: When you are doing calculations using vectors, remember to treat one direction as the positive direction; the force in this diagram will produce a deceleration

Investigating acceleration

Fig. 1b.05 shows a set of apparatus that can be used to investigate the link between force, mass and acceleration.

The slope of the ramp is adjusted so that the trolley just rolls down it when pushed. The force of gravity pulling the trolley down the ramp just balances the frictional forces that are trying to slow the trolley down. This means that you can ignore the effects of friction on your investigation.

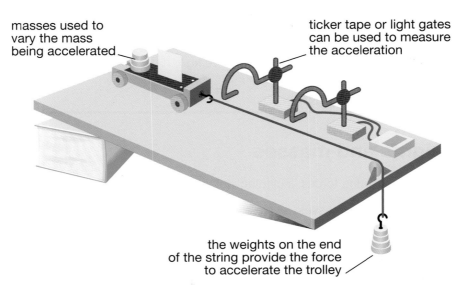

masses used to vary the mass being accelerated

ticker tape or light gates can be used to measure the acceleration

the weights on the end of the string provide the force to accelerate the trolley

Fig. 1b.05: Apparatus for investigating acceleration

Weight, mass and gravity

Weight is a force produced by the gravitational attraction between two masses. On Earth, gravity attracts a mass of 1 kilogram with a force of 9.81 N (although you are often told to use a value of 10 N/kg in exams).

We use the symbol g for the **gravitational field strength**.

Weight, mass and gravity are related by this formula:

$$weight = mass \times g$$
$$W = m \times g$$

You can rearrange this formula to calculate mass or g:

$$mass = \frac{weight}{g}$$
$$g = \frac{weight}{mass}$$

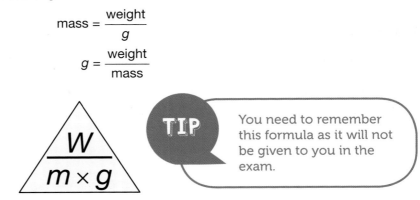

TIP You need to remember this formula as it will not be given to you in the exam.

Fig. 1b.06: Formula triangle for weight, mass and g

The gravitational field strength is different on different planets and moons. The gravitational field strength on the Moon is less than on the Earth because the Moon has a smaller mass.

Worked example

A 10 kg mass has a weight of 16 N on the Moon. What is the gravitational field strength on the Moon?

Answer

$$g = \frac{W}{m}$$

$$= \frac{16 \text{ N}}{10 \text{ kg}}$$

$$= 1.6 \text{ N/kg}$$

Measuring weights and masses

A spring balance is used for measuring forces, including weights. Weight is always measured in newtons.

We can also use balances to compare masses: as we always weigh things in the Earth's gravitational field, the mass of an object can be worked out from its weight.

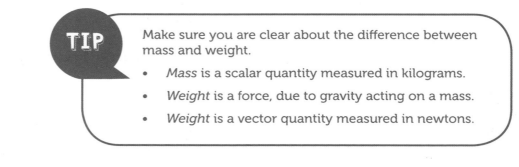

TIP Make sure you are clear about the difference between mass and weight.

- *Mass* is a scalar quantity measured in kilograms.
- *Weight* is a force, due to gravity acting on a mass.
- *Weight* is a vector quantity measured in newtons.

Worked example

What is the mass of an object that weighs 5 N? Assume that *g* has a value of 10 N/kg.

Answer

$$\text{mass} = \frac{\text{weight}}{g}$$

$$= \frac{5 \text{ N}}{10 \text{ N/kg}}$$

$$= 0.5 \text{ kg}$$

Falling objects and terminal velocity

If you drop an object it falls downwards.

If you do this in a vacuum on the Earth, the object will accelerate at 9.81 m/s^2, whatever its mass. A greater mass has a greater weight, which means that there is a greater force accelerating it. However, a greater mass needs a greater force to give it a certain acceleration. These two effects balance each other out, so that falling objects always accelerate at 9.81 m/s^2, as long as there are no other forces on them.

Objects falling in air (or water) do not fall at a constant speed because they are affected by friction (see Fig. 1b.07).

- The faster an object is moving through air or water, the greater the friction.
- The acceleration of a falling object is initially 9.81 m/s^2, but as it gains speed the air resistance increases.
- So the resultant downwards force decreases.
- The acceleration gradually decreases to zero.
- Once the acceleration has reached zero the velocity stops changing.
- The falling object has reached its **terminal velocity**.

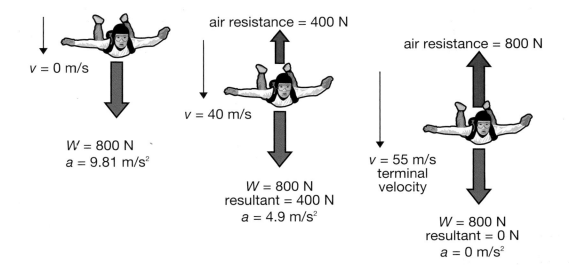

Fig 1b.07: A skydiver reaches terminal velocity

> **TIP**
>
> **Units for g**
> You may see the approximate value for g given as the gravitational field strength (10 N/kg) or as an acceleration (10 m/s^2). These are equivalent.

Investigating falling objects

It takes a skydiver around 10 seconds to reach terminal velocity, during which time they have fallen around 500 metres. Lighter objects, such as paper cake cases, will reach terminal velocity in a much shorter time, but will also be more affected by air currents.

This means that it is not very practical to investigate the factors affecting the speed of objects falling in air in a school laboratory. Investigations into falling objects can be carried out using denser fluids such as oil or wallpaper paste. Fig. 1b.08 shows a possible set of apparatus that can be used for this.

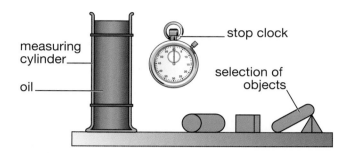

Fig. 1b.08: Apparatus for investigating falling objects

Stopping distances

The **stopping distance** of a vehicle is the distance it travels along the road while it is slowing down to a stop. This distance is often split up into:

- thinking distance
- braking distance.

The **thinking distance** is the distance the vehicle travels while the driver is deciding to stop. It is increased if:

- the driver is tired or has been drinking alcohol, as this will slow down their reaction time and increase the distance the vehicle moves while they are deciding to stop
- the vehicle is going faster, as it will move further in the same reaction time.

The **braking distance** is the distance the vehicle travels while the brakes are bringing it to a stop. The braking distance will increase if:

- the vehicle is moving faster, as the deceleration provided by the brakes will take longer to reduce the speed
- the mass of the vehicle is greater, as the force from the brakes will provide a smaller deceleration
- the road is wet or slippery, as the wheels may skid due to reduced friction
- the tyres are worn, as they may skid due to reduced friction
- the brakes are worn, as they will not work as well as they should.

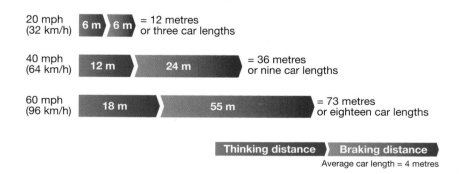

Fig. 1b.09: This stopping distance chart, similar to that used in the UK Highway Code, shows the thinking and braking distances when travelling at different speeds

Momentum

Momentum is the property of a moving body that tends to keep it moving. It depends on the mass and the velocity.

Momentum, mass and velocity are related by this formula:

$$\text{momentum} = \text{mass} \times \text{velocity}$$
$$p = m \times v$$

The units for momentum are kilogram metres per second (kg m/s). The formula can be rearranged to calculate mass or velocity:

$$\text{mass} = \frac{\text{momentum}}{\text{velocity}}$$

$$\text{velocity} = \frac{\text{momentum}}{\text{mass}}$$

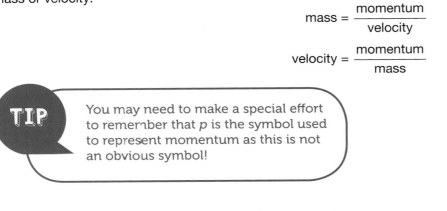

TIP You may need to make a special effort to remember that *p* is the symbol used to represent momentum as this is not an obvious symbol!

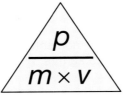

Fig. 1b.10: Formula triangle for momentum, mass and velocity

Worked example

What is the momentum of a 2 kg bowling ball moving at 1.5 m/s?

Answer

$$\text{momentum} = \text{mass} \times \text{velocity}$$
$$= 2 \text{ kg} \times 1.5 \text{ m/s}$$
$$= 3 \text{ kg m/s}$$

Conservation of momentum

Momentum is **conserved**, which means that it is the same before and after an event, as long as no energy is added to or removed from the system.

Fig. 1b.11 shows two trolleys held together with a compressed spring between them. When the trolleys are released the spring pushes them apart. This system initially had zero momentum because the trolleys were not moving. This means that the final momentum must also be zero. We can use this fact to calculate the velocity of trolley Y after the trolleys are released.

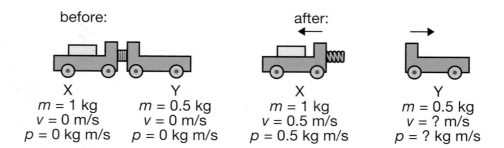

before:		after:	
X	Y	X	Y
$m = 1$ kg	$m = 0.5$ kg	$m = 1$ kg	$m = 0.5$ kg
$v = 0$ m/s	$v = 0$ m/s	$v = 0.5$ m/s	$v = ?$ m/s
$p = 0$ kg m/s	$p = 0$ kg m/s	$p = 0.5$ kg m/s	$p = ?$ kg m/s

Fig. 1b.11: The total momentum is zero before the trolleys are released, so it must be zero afterwards

Worked example

After the trolleys in Fig.1b.11 are released, the momentum of X is 0.5 kg m/s. As the total momentum is zero, the momentum of trolley Y must be −0.5 kg m/s. What is the velocity of trolley Y?

Answer

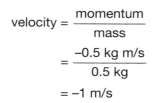

$$\text{velocity} = \frac{\text{momentum}}{\text{mass}}$$

$$= \frac{-0.5 \text{ kg m/s}}{0.5 \text{ kg}}$$

$$= -1 \text{ m/s}$$

The minus sign indicates that trolley Y is moving in the opposite direction to trolley X.

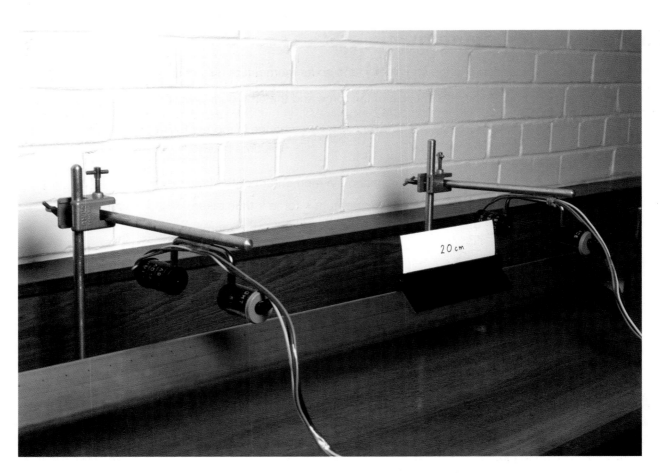

Fig. 1b.12: A linear air track with light gates

A linear air track (Fig. 1b.12) can be used to investigate momentum and collisions. Air is blown upwards through small holes along the track. This allows the gliders to move with almost zero friction.

Worked example

A 0.5 kg glider is stationary on an air track. It is hit by a 0.3 kg glider moving at 0.8 m/s, and the two gliders then stick together. What is the velocity of the combined gliders after the collision?

Answer

$$\text{momentum before collision} = 0.5 \text{ kg} \times 0 \text{ m/s} + 0.3 \text{ kg} \times 0.8 \text{ m/s}$$

$$= 0.24 \text{ kg m/s}$$

$$\text{momentum after collision} = 0.24 \text{ kg m/s}$$

$$\text{velocity after collision} = \frac{0.24 \text{ kg m/s}}{(0.5 \text{ kg} + 0.3 \text{ kg})}$$

$$= 0.3 \text{ m/s}$$

Changing the momentum

A force applied to an object can change its velocity, and so its momentum also changes.

The change in momentum depends on the size of the force and the time during which the force is applied. These factors are related by the following formula:

$$\text{force} = \frac{\text{change in momentum}}{\text{time taken}}$$

This formula can be rearranged to calculate the change in momentum or the time taken:

$$\text{change in momentum} = \text{force} \times \text{time taken}$$

$$\text{time taken} = \frac{\text{change in momentum}}{\text{force}}$$

Fig. 1b.13: Formula triangle for the change in momentum

Worked example

A 1000 kg car travelling at 20 m/s comes to a stop in 2 seconds. What force is needed to achieve this?

Answer

$$\text{momentum of moving car} = \text{mass} \times \text{velocity}$$
$$= 1000 \text{ kg} \times 20 \text{ m/s}$$
$$= 20\,000 \text{ kg m/s}$$

As the final momentum is zero, the change in momentum is also 20 000 kg m/s.

$$\text{force} = \frac{\text{change in momentum}}{\text{time taken}}$$
$$= \frac{20\,000 \text{ kg m/s}}{2 \text{ s}}$$
$$= 10\,000 \text{ N}$$

Car safety features

As the example on page 24 shows, the forces on a car in a crash can be very large. This force is also applied to the passengers.

Cars have various safety features to help the occupants to survive a crash. Many of these safety features are designed to allow the occupants to come to a stop over a longer time. This reduces the size of the force acting on them.

- **Crumple zones** in the front of the car get crushed if the car hits something, which increases the time the car takes to stop.
- **Seat belts** hold the passengers into the car, so they also get the benefit of the crumple zone.
- **Air bags** increase the time it takes a person's head to come to a stop compared to hitting their head on the steering wheel or dashboard.

Newton's third law

Sir Isaac Newton made three important observations about the way that objects move. The first of these laws says that moving objects continue to move in a straight line unless there is a force acting on them, and the second explains how the acceleration of a body depends on its mass and on the force applied to it.

Newton's third law is often stated as:

- Every action has an equal and opposite reaction.

This means that if you push on something, it pushes back on you with a force equal in size but opposite in direction. When you stand on the floor, your weight is pushing down on the floor. The floor is pushing back up on your feet with an equal force – if it wasn't, you would fall through the floor!

If you stand in front of someone and push them away, it is not always obvious that they are also exerting a force on you. However, if you tried this when you were both wearing ice skates, or both sitting on chairs with wheels, you would move away from each other.

TIP

Don't get balanced forces and action/reaction forces confused. Balanced forces are forces *on the same object* that are equal and opposite in direction. They cancel each other out and the object has a zero resultant force acting on it.

Action/reaction forces are also equal and opposite, but they act *on different objects*.

Moments

A turning force is called a **moment**.

- The moment of a force depends on the size of the force and the distance between the force and a pivot.
- The distance must be measured perpendicular to the line of action of the force.

The formula below is used to calculate the moment of a force. The force must be measured in newtons and the distance in metres. The units for moments are newton metres (N m).

$$\text{moment} = \text{force} \times \text{perpendicular distance from pivot}$$
$$M = f \times d$$

moment = 4.8 N m

Fig. 1b.14: The distance used to calculate a moment must be measured perpendicular to the force

The formula can be rearranged to allow you to calculate force or distance:

$$\text{force} = \frac{\text{moment}}{\text{perpendicular distance}}$$

$$\text{perpendicular distance} = \frac{\text{moment}}{\text{force}}$$

TIP You need to remember this formula as it will not be given to you in the exam.

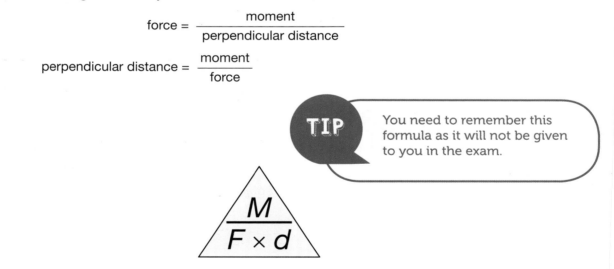

Fig. 1b.15: Formula triangle for moment, force and distance

The idea of moments explains why it is easier to undo a stiff nut with a long spanner than with a short one, and why tools that gardeners use to cut through branches have much longer handles than the ones they use to cut flowers or small twigs. The longer the handle, the greater the turning force about the pivot for a given force applied to the end of the handle.

Moments and balancing

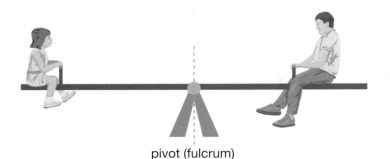

pivot (fulcrum)

Fig. 1b.16: Two people of different weights can balance on a seesaw by sitting at different distances from the pivot

The seesaw in Fig. 1b.16 is balanced. The moment from the boy is trying to turn the seesaw in a clockwise direction. This moment is being balanced by the moment from the girl, which is trying to move it in an anticlockwise direction. The seesaw is in **equilibrium**.

The **principle of moments** says that when an object is in equilibrium the sum of the clockwise moments about a point is equal to the sum of the anticlockwise moments about the point.

Worked example

The boy in Fig. 1b.16 has a weight of 600 N and is sitting 1.5 m from the pivot. The girl has a weight of 450 N. How far from the pivot must she sit to balance the seesaw?

Answer

clockwise moment = anticlockwise moment

600 N × 1.5 m = 450 N × ? m

$$\text{distance} = \frac{900 \text{ N m}}{450 \text{ N}}$$

$$= 2 \text{ m}$$

Fig. 1b.17: You can demonstrate the principle of moments using a metre rule, a pivot and some masses; the product of the weight and the distance will be equal on each side of the pivot

Forces on a beam

You can use the principle of moments to work out the forces needed to support the ends of a light beam. 'Light' means that you do not need to take the mass of the beam itself into account.

If an object is placed in the middle of the beam, the force on each support will be half the weight of the object. If the object is not placed in the middle, there will be a greater force on the support closest to the object.

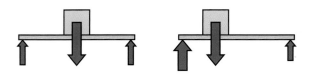

Fig. 1b.18: Forces on a light beam

Centre of gravity

The **centre of gravity** (sometimes called **centre of mass**) is the point in a body where all the weight can be thought to act.

For a symmetrical body, the centre of gravity is on a line of symmetry.

CAM

A suspended body will always hang with its centre of gravity below the point of suspension. You can use this idea to find the centre of gravity of an irregular sheet (see Fig. 1b.19).

(a) Suspend the shape from a pin and use a plumb line to mark the vertical on it.

(b) Suspend the shape from a different point and mark the vertical again. The point where the two lines cross is the centre of gravity of the shape.

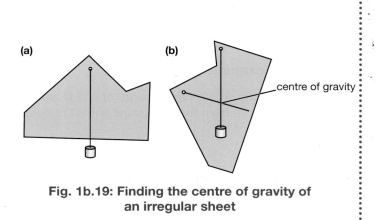

(a)　　　(b)

centre of gravity

Fig. 1b.19: Finding the centre of gravity of an irregular sheet

Centre of mass and stability

A stable object is one that returns to its original position if it is disturbed. A book lying flat on the table is stable. A book standing on its end is not very stable, as it only takes a small push to knock it over.

The stability of an object depends on the position of its centre of mass and the width of its base. The weight of an object acts downwards from its centre of mass.

If the line of action of the weight is still within the area of the base when the object is tilted, it will return to its original position. The lower the centre of mass and the wider the base, the more stable the object is.

line of action of weight

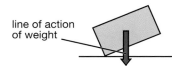

stable object: this will return to its original position

unstable object: this will fall over

Fig. 1b.20: The stability of an object depends on the position of its centre of mass

Force and extension

A force applied to a spring will make it longer.

The effect of forces on springs can be investigated by hanging different weights on a spring and measuring its length.

The **extension** of the spring is the stretched length minus the original length.

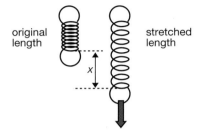

original length

stretched length

x

Fig. 1b.21: Stretching helical (coiled) springs; the extension of the spring is the stretched length minus the original length, as shown by x on this diagram

TIP

Remember that when answering questions about springs it is almost always the extension that is involved, not the length.

A wire stretches in a similar way to a spring. Fig. 1b.22 shows how the extension varies with the applied force for springs and wires.

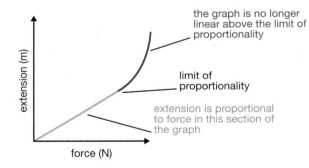

the graph is no longer linear above the limit of proportionality

limit of proportionality

extension is proportional to force in this section of the graph

extension (m)

force (N)

Fig. 1b.22: Extension versus force for a spring; the force on the spring is sometimes referred to as the *load*

For the **linear** section of the graph (shown in blue) the spring is following **Hooke's law**, which states that the extension is proportional to the load. The spring behaves **elastically**. This means that if the force is removed, the spring will return to its original length.

If the force stretches the spring beyond the **limit of proportionality**, the extension is no longer proportional to the load (shown in red). The spring has exceeded its **elastic limit**. If the force is removed, the spring will not return to its original length but will be permanently stretched. It is no longer behaving elastically.

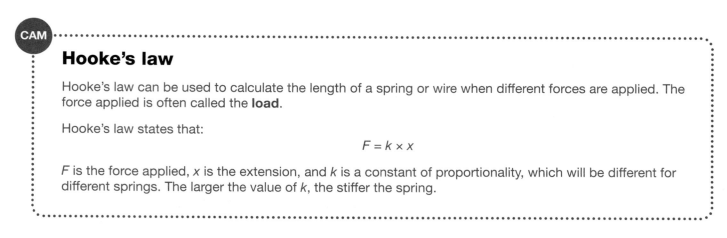

Hooke's law

Hooke's law can be used to calculate the length of a spring or wire when different forces are applied. The force applied is often called the **load**.

Hooke's law states that:

$$F = k \times x$$

F is the force applied, x is the extension, and k is a constant of proportionality, which will be different for different springs. The larger the value of k, the stiffer the spring.

Stretching rubber bands

When you stretch a rubber band, the extension is not proportional to the force applied.

The rubber is hard to stretch at first and then becomes easier. This is shown by the gradient becoming steeper on the graph, which means that you are getting a greater increase in the extension for a certain increase in the force.

Eventually the rubber becomes stiffer and harder to stretch again.

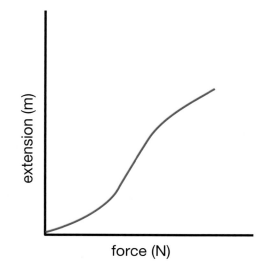

Fig. 1b.23: Extension versus force for a rubber band

You should be able to:

- ★ recall and use the names and symbols of the units for the following quantities: mass, momentum, acceleration, weight, velocity (see page 14)
- ★ identify different types of force, including friction and weight (see page 14)
- ★ explain the difference between vector and scalar quantities (see page 14)
- ★ add forces that act along a line (see page 15)
- ★ **find the resultant of two forces that do not act along the same line (see page 15)**
- ★ recall and use the formula relating force, mass and acceleration, rearranging it if necessary (see page 16)
- ★ recall and use the formula relating weight, mass and the acceleration due to gravity, rearranging it if necessary (see page 17)
- ★ describe the forces acting on falling objects and explain why falling objects reach a terminal velocity (see page 19)
- ★ describe the factors affecting vehicle stopping distances (see page 20)
- ★ recall and use the formula relating momentum, mass and velocity, rearranging it if necessary (see page 21)
- ★ use conservation of momentum to calculate the mass, momentum or velocity of objects (see page 22)
- ★ recall and use the formula relating force, change in momentum and time taken, rearranging it if necessary (see page 24)
- ★ use ideas about momentum to explain car safety features such as air bags, crumple zones and seat belts (see page 25)
- ★ name some pairs of action and reaction forces, such as the ones acting on you when you are sitting in a chair (see page 25)
- ★ recall and use the formula for calculating moments (see page 26)
- ★ use the principle of moments to calculate forces on objects in equilibrium (see page 27)
- ★ explain the centre of gravity of a body (see page 28)
- ★ explain how to find the centre of mass of an irregular shape (see page 28)
- ★ describe how the position of the centre of mass affects the stability of an object (see page 28)
- ★ explain the meanings of elastic and limit of proportionality (see page 29)
- ★ sketch load–extension graphs for springs, wires and rubber bands (see page 29)
- ★ recall Hooke's law (see page 30).

Review questions

1. Name two different friction forces.

2. What are the resultant forces on the boxes below?

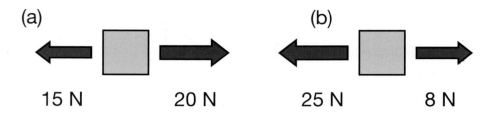

(a)   15 N   20 N   (b)   25 N   8 N

CAM 3. A 50 N force is acting at right angles to a 20 N force. Draw a scale drawing to work out the size and direction of the resultant force.

4. (a) What acceleration is produced by a 10 N force acting on a 2 kg mass?

   (b) An object accelerates at 5 m/s^2 when a force of 15 N acts on it. What is its mass?

5. What is the weight of a 10 kg mass?

6. (a) What is the acceleration of a body that has reached terminal velocity?

   (b) How does the acceleration of a skydiver change between the time they jump out of the aeroplane and the time they reach terminal velocity?

   (c) Why does the acceleration change?

7. What is the momentum of a 20 kg mass travelling at 50 m/s?

8. A 15 kg mass has a momentum of 75 kg m/s. What is its velocity?

9. A moving trolley hits a stationary one and they stick together. What can you say about the velocity of the combined trolley after the collision?

10. A moving body changes its momentum by 20 kg m/s in a time of 4 seconds. What was the force acting on it?

11. Name two safety features in cars that are designed to reduce the force on passengers in a collision.

12. Why is it easier to undo a stiff nut with a long spanner than a short one?

13. A boy weighing 400 N sits 2 m from the pivot on a seesaw. Where does a 500 N girl have to sit so that the seesaw will balance?

14. A plank is making a simple bridge over a stream. Explain how the forces on the stream banks will change as you walk across the plank.

CAM 15. Describe two features of a Bunsen burner that help to make it stable, and explain why they work.

Practice questions

1. The UK Highway Code recommends that cars on a motorway leave at least a 2-second gap from the car in front. This means that the length of the gap will be different for different speeds.

    (a) Why does the length of the recommended gap depend on the speed of the car? **(3)**

    (b) The Highway Code divides the stopping distance into two parts, each affected by different factors. Name the two parts. **(1)**

    (c) This is a speed limit sign from a French motorway.

    The speeds are in km/h. Suggest how the weather conditions shown could affect the two distances you named in part (b). **(2)**

2. A skydiver and her parachute have a total mass of 70 kg and a weight of 700 N. The air resistance acting on her at one point in freefall is 500 N.

    (a) What is the resultant force on her at this point? State the size of the force and its direction. **(2)**

    (b) What is her acceleration? **(4)**

    (c) What is the upward force on her when she has reached terminal velocity? Explain your answer. **(2)**

    (d) Explain how her terminal velocity in freefall would be different if she had a larger mass. **(2)**

    (e) When she opens her parachute, her velocity changes from 55 m/s to 5 m/s in 3 seconds.

    　　(i) Calculate her acceleration. **(3)**
    　　(ii) Calculate the force on her while her parachute is opening. **(2)**

3. A car is moving at 30 m/s. It has a mass of 1500 kg.

    (a) What is its momentum? **(3)**

    (b) The car crashes into a stationary car with a mass of 1000 kg and the two cars move along the road together. What is the speed of the two cars after the collision? **(3)**

    (c) What assumption did you make in your answer to part (b)? **(1)**

4. A 0.15 kg ball moving at 6 m/s bounces off a wall. It has the same speed after the collision as before.

    What force does the wall exert on the ball, if it is in contact with the ball for 0.005 seconds? Show all your working. **(4)**

5. A student is investigating the effect of force on acceleration. The apparatus he uses is shown in Fig. 1b.05 on page 17.

He changes the force by removing masses from the trolley one at a time and adding them to the end of the string. He measures the acceleration for each different force.

(a) What is the force pulling the trolley when the total mass on the string is 1.2 kg? Assume *g* = 10 N/kg. **(2)**

(b) Explain how the light gates and a datalogger measure the speed of the trolley. **(2)**

(c) How can the student use information from the light gates to measure the acceleration? **(3)**

(d) Why does the ramp have a slight slope? **(2)**

(e) Explain why the student moves the masses from the trolley to the end of the string, instead of just adding masses to the end of the string from a pile on the table. **(2)**

6. The drawing shows a trapdoor which is lifted by a vertical rope.

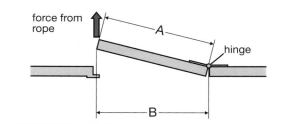

(a) Explain which distance you should use when calculating the moment of the force in the rope about the hinge. **(2)**

The drawing shows a bridge over a canal. The bridge can be lifted by hand by pulling on the rope marked A.

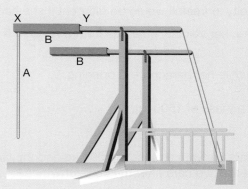

(b) The parts labelled B are made of wood covered in iron to make them heavier. Explain why this is done. **(2)**

(c) Explain why rope A is fastened at point X rather than point Y. **(2)**

CAM **7.** The drawing shows two glasses.

tumbler   stemmed glass

(a) Explain why the stemmed glass has a wide base. (2)

(b) How does filling the stemmed glass affect its stability? Explain your answer. (3)

8. The drawing shows a bolt used for fastening a gate. There is a spring to make sure the bolt stays in the closed position unless someone pulls on the end. The spring stretches as the bolt is pulled back.

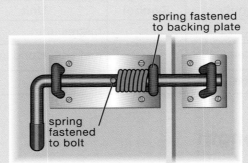

spring fastened
to backing plate

spring
fastened
to bolt

(a) Explain why it gets harder to pull the bolt as you pull it further. (1)

(b) Explain why the spring must not exceed its limit of proportionality. (3)

C Astronomy

Units

You will use the following quantities and units in this section. You should already know most of these from your previous studies.

Quantity	Symbol	Unit
radius	r	metres (m)
orbital speed	v	metres/second (m/s)
time period	T	seconds (s)

Gravitational field strength

The **gravitational field strength** of a body is the gravitational force it exerts on every kilogram of matter. It is represented by the symbol g.

The gravitational field strength of a body depends on its mass. The gravitational field strength on the surface of a body also depends on the radius of the body, as the field strength gets weaker as you get further from its centre. This means that the gravitational field strength at the surface is different for all the different bodies in the solar system. The Earth's gravitational field strength is approximately 9.81 N/kg at the surface.

TIP You have already met g in Section 1B, in this formula: weight = mass × g.

Physics A* Study Guide

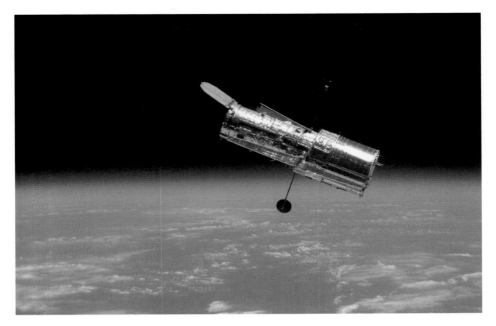

Fig. 1c.01: The Hubble Space Telescope is about 559 km from the Earth; the force of gravity on every kilogram of the Hubble Space Telescope is only about 8.3 N

Gravity and orbits

The Earth is part of the solar system. The Earth and the other planets **orbit** the Sun. The gravitational force of the Sun keeps all these bodies in orbit around the Sun.

The Moon is a **natural satellite** of the Earth. There are many **artificial satellites** orbiting around the Earth. These include scientific satellites such as the Hubble Space Telescope, satellites used for broadcasting TV programmes or telephone calls, and satellites used for photographing the Earth or helping to forecast the weather.

The Moon and the artificial satellites are kept in their orbits by the gravitational force of the Earth. Other planets in our solar system also have **moons** orbiting them.

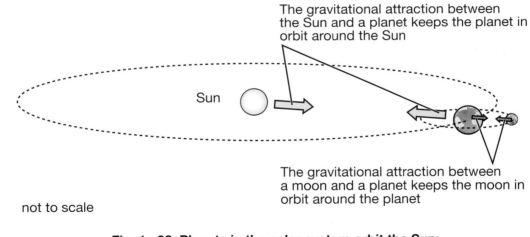

The gravitational attraction between the Sun and a planet keeps the planet in orbit around the Sun

Sun

not to scale

The gravitational attraction between a moon and a planet keeps the moon in orbit around the planet

Fig. 1c.02: Planets in the solar system orbit the Sun; moons are smaller bodies that orbit planets

The solar system is part of the Milky Way **galaxy**. Galaxies are large collections of billions of stars, and there are billions of galaxies in the **universe**.

Orbital speeds

The planets all orbit the Sun in approximately circular orbits. The length of time it takes a planet to orbit the Sun once is its **orbital period (T)**.

The speed of a planet is related to its orbital period and the radius of its orbit by this formula:

$$\text{orbital speed} = \frac{2 \times \pi \times \text{orbital radius}}{\text{time period}}$$

$$v = \frac{2 \times \pi \times r}{T}$$

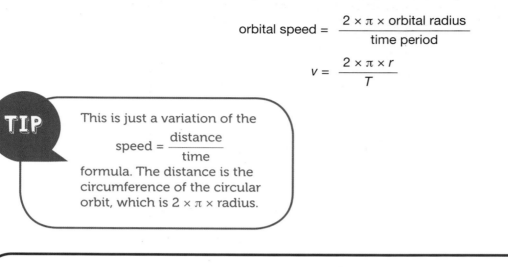

TIP This is just a variation of the
$$\text{speed} = \frac{\text{distance}}{\text{time}}$$
formula. The distance is the circumference of the circular orbit, which is $2 \times \pi \times$ radius.

Worked example

The Earth is approximately 150 000 000 km from the Sun. What is its speed?

Answer

The speed will be in metres per second, so the distance must be in metres.
150 000 000 km = 150 000 000 000 m (or 1.5×10^{11} m)

The time must be in seconds. A year is approximately $365 \times 24 \times 60 \times 60 = 31\ 536\ 000$ seconds.

$$\text{orbital speed} = \frac{2 \times \pi \times 1.5 \times 10^{11}}{31\ 536\ 000\ \text{s}}$$

$$= 29\ 886\ \text{m/s}$$

$$= 30\ 000\ \text{m/s (or } 3 \times 10^4\ \text{m/s)}$$

Comets

A **comet** is a ball of dirty ice orbiting the Sun. Unlike the planets, comets have a very elliptical orbit, with the Sun at one focus of the **ellipse**. Most comets come from the outer solar system and can have periods from a few years to many thousands of years.

The speed of a comet varies as it moves around its orbit. The comet moves fastest when it is closest to the Sun, and moves much more slowly in the distant part of its orbit.

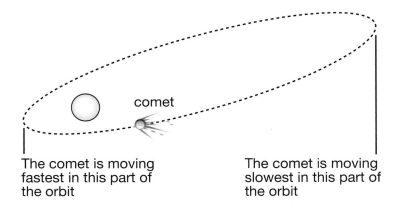

The comet is moving fastest in this part of the orbit

The comet is moving slowest in this part of the orbit

Fig. 1c.03: The orbits of comets are elliptical

TIP

Check your spelling and handwriting if you are writing about comets. If the examiner thinks you have written 'eclipse' instead of ellipse you will not get the mark.

CAM

Motion in curved paths

A moving object will continue to move in a straight line at a constant speed unless it is acted on by a force. This means that an object moving on a curved path must have a force acting on it to change its direction. For a circular path, this **centripetal force** acts towards the centre of the circle.

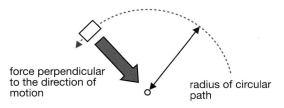

force perpendicular to the direction of motion

radius of circular path

Fig. 1c.04: An object moving in a circle has a centripetal force acting towards the centre of the circle

The size of the force needed to keep an object moving on a curved path depends on the speed of the object, its mass, and the radius of the curve.

- The larger the mass of the object, the larger the force needed.
- The higher the speed of the object, the larger the force needed.
- The smaller the radius of the circle, the larger the force needed.

You should now be able to:

★ explain what gravitational field strength is and that it is different on different bodies in the solar system (see page 36)

★ explain the effects of gravitational force in the solar system (see page 37)

★ explain what moons are (see page 37)

★ describe what a galaxy is (see page 37)

★ use the formula that relates orbital speed, radius and time period (see page 38)

★ describe the differences between the orbits of planets and comets (see page 39)

CAM ★ describe the factors that affect the size of a centripetal force (see page 39).

Review questions

1. What do these bodies orbit?

    (a) planet

    (b) moon

    (c) comet

2. The Moon is 384 399 km from the centre of the Earth, and takes 27.3 days to complete an orbit. How fast is it moving?

3. Mercury travels around the Sun at approximately 48 000 m/s. Its average distance from the Sun is 5.79×10^{10} m. How long does it take to complete one orbit?

Practice questions

1. Titan is one of the moons of Saturn. The Huygens spacecraft landed on Titan in 2005. The Huygens lander had a mass of 319 kg. Its weight on Titan was 431 N.

    (a) What is the gravitational field strength on Titan? **(2)**

    (b) If Huygens had been taken to the Moon instead, its weight would have been 517 N. What does this tell you about Titan and the Moon? **(2)**

    (c) Titan orbits Saturn once every 5.9 days. The radius of its orbit is approximately 355 000 km. What is its orbital speed, in m/s? **(4)**

Section Two

A Mains electricity

You will be expected to:

* ★ recall some of the hazards of electricity
* ★ describe how insulation is used in domestic appliances
* ★ explain how earthing, fuses and circuit breakers are used
* ★ describe some ways in which electrical heating is used
* ★ explain that a current in a resistor transfers energy
* ★ recall and use the formula relating power, current and voltage, including rearranging it if necessary
* ★ use the formula relating energy, current, voltage and time, including rearranging it if necessary
* ★ explain the difference between alternating current and direct current.

Units

You will use the following quantities and units in this section. You should already know most of these from your previous studies.

Quantity	Symbol	Unit
current	I	amperes (or amps) (A)
energy	E	joules (J)
power	P	watts (W)
time	t	seconds (s)
voltage/potential difference	V	volts (V)

You also need to recall and use the standard symbols for electrical components. A full list of these is given in Appendix 2 (on page 242).

Dangers of mains electricity

Electricity is a very convenient way of transferring energy to appliances that can do useful work, such as motors or computers. Electricity can be hazardous if it is not used properly. Hazards include fires, and electrical shocks leading to severe burns or even death.

Some of the rules for using electricity safely are listed below.

- Replace any frayed cables, as damaged insulation could cause an electrical shock or a fire if something comes into contact with the bare wires.
- Replace any damaged plugs, as damage may expose live wires or connections, and could cause electric shocks.
- Avoid water around sockets, and do not touch sockets with wet hands. Water can conduct electricity if the voltage is high enough, so water near electricity could lead to electrical shocks.
- Avoid long cables, especially on kitchen appliances, because they can be a trip hazard and could lead to water spills.
- Never poke metal objects into sockets. The metal object will conduct electricity and you will get an electric shock.

Insulation

The part of an electrical cable that carries the current is made of metal, because metals **conduct** electricity.

The wire is surrounded by plastic, because plastic is an insulating material. The casings of plugs are also made out of plastic. The plastic **insulation** stops people coming into contact with the conducting part of the circuit, so it prevents electric shocks.

Many appliances have three wires in their cables: live, neutral and earth wires. The live and neutral wires form part of the circuit that allows current to flow through the appliance. The earth wire is there for safety (see page 44).

Some appliances that are made of insulating materials, such as plastic kettles or hairdryers, do not need earth wires. Appliances with insulating cases are **double insulated**.

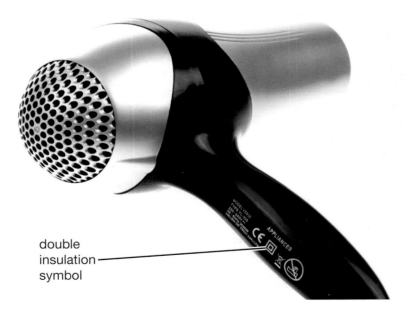

double insulation symbol

Fig. 2a.01: Appliances with double insulation have this symbol on them

Earthing, fuses and circuit breakers

Most plugs contain **fuses**. A fuse is a short piece of wire enclosed in a cylinder. The wire is designed to melt and break the circuit if the current is too high. This can prevent damage to the appliance. The fuse must be replaced once the fault has been corrected.

Fuses are available in three ratings. In the UK these are 3 A, 5 A and 13 A. A fuse with a 3 A rating will melt if the current is greater than 3 A.

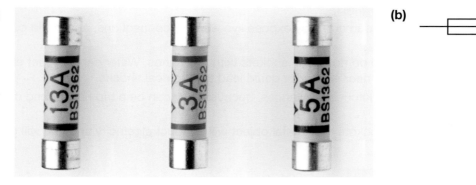

(a)

(b)

Fig. 2a.02: (a) These fuses are used in plugs to protect the appliance against high currents
(b) The electrical circuit symbol for a fuse

Faults in appliances can sometimes lead to the **live wire** coming into contact with the metal outer casing. This may happen in something like a washing machine. If someone touches the casing of the washing machine, the current can flow through them to earth. The current through the person is likely to be high enough to make the fuse melt and cut off the current, but not before the person has received an electric shock (see Fig. 2a.03(a)).

If the appliance has an **earth wire**, a fault that makes the casing live would result in the current flowing through the earth wire to earth. The earth wire provides a low-resistance path for the current, and so a high current will flow (Fig. 2a.03(b)). This current will be high enough to melt the fuse. The appliance is therefore disconnected from the mains supply as soon as the fault occurs, and anyone touching it later will not receive an electric shock (Fig. 2a.03(c)).

(a)

live wire
touching
the casing

fuse
live wire
neutral wire

(b)

current flows
through the
case and the
earth wire
fuse
live wire
neutral wire
earth wire

(c)

live wire
neutral wire
earth wire

fuse

Fig. 2a.03: In a faulty appliance with a metal casing:
(a) With no earth wire a person could receive an electric shock before the fuse melts
(b) With an earth wire the current will flow to earth, melting the fuse so the current is cut off as soon as the fault occurs
(c) With an earth wire a person will not get a shock if they touch the faulty appliance

Circuit breakers are also used to cut off a current if it becomes too high. They often work using electromagnets which open a set of contacts when the current exceeds a certain value. They can be reset using a switch after the fault has been corrected.

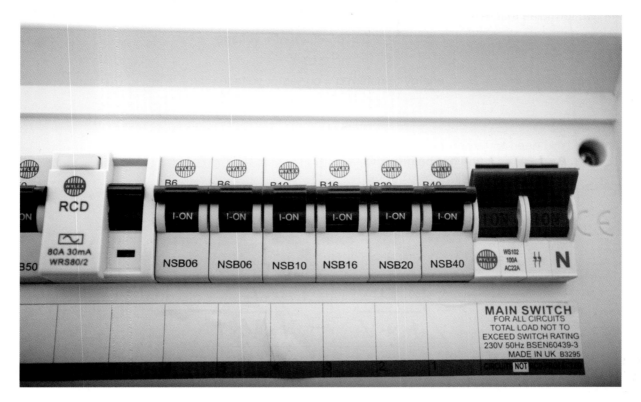

Fig. 2a.04: Circuit breakers are used in the consumer unit, where mains electricity enters the house

Electricity for heating

Electricity is a way of transferring energy. All components have some resistance to the flow of electricity (see Section 2B). When a current flows, energy is transferred to the wires and they heat up. The higher the resistance, the more energy is transferred by heating.

Electric fires have a coil of high resistance wire. This transfers energy by heating when electricity flows through it, and the temperature of the wire rises. Heat energy from the coil is transferred to the room. Heating coils are also used in fan heaters, where a fan blows the hot air into the room, and in kettles, tumble dryers, washing machines, electric cookers and toasters.

Electricity and power

Power is the amount of energy transferred each second. The units for power are **watts** (W).
There are two formulae that relate power, current and voltage:

$$\text{power} = \text{current} \times \text{voltage}$$
$$P = I \times V$$

$$\text{energy transferred} = \text{current} \times \text{voltage} \times \text{time}$$
$$E = I \times V \times t$$

> **TIP**
> You need to remember the formula $P = I \times V$ as it will not be given to you in the exam. You need to be able to rearrange both formulae to calculate different quantities.

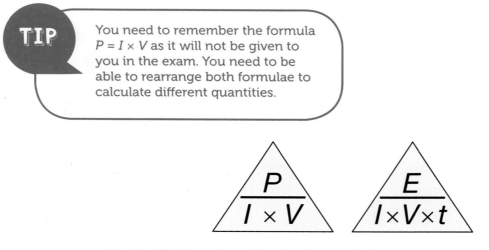

Fig. 2a.05: These formula triangles can help you to rearrange the formulae

> **TIP**
> The formula for energy transferred can also be written as $E = P \times t$, because $V \times I$ gives the power.

Worked example

A 100 W light bulb uses a voltage of 230 V. What current flows through it?

Answer

First choose the correct formula. The question gives the power and voltage, and asks about current. It does not mention energy or time, so you need to rearrange $P = I \times V$.

$$\text{current} = \frac{\text{power}}{\text{voltage}}$$

$$= \frac{100 \text{ W}}{230 \text{ V}}$$

$$= 0.43 \text{ A}$$

Worked example

A light bulb is left on for 5 hours. How much energy does it transfer in that time? The bulb uses the 230 V mains supply, and draws a current of 0.09 A.

Answer

The question gives you power, voltage, time and current. You need to use the formula $E = V \times I \times t$.

Time must be in seconds:
$$\text{time} = 5 \times 60 \times 60$$
$$= 18\ 000\ \text{s}$$

$$\text{energy} = \text{voltage} \times \text{current} \times \text{time}$$
$$= 230\ \text{V} \times 0.09\ \text{A} \times 18\ 000\ \text{s}$$
$$= 372\ 600\ \text{J}$$

Choosing fuses

The formula for power, voltage and current can be used to work out the correct fuse to use for an appliance. You need to calculate the current that the appliance should use when it is working normally, and to choose the fuse with a rating just above this current.

Worked example

A bedside light with a 20 W light bulb is plugged into the 230 V mains supply. Which fuse should be used in the plug: 3 A, 5 A or 13 A?

Answer

$$\text{current} = \frac{\text{power}}{\text{voltage}}$$
$$= \frac{20\ \text{W}}{230\ \text{V}}$$
$$= 0.09\ \text{A}$$

A 3 A fuse should be used.

In the example above you could use a 13 A fuse in the plug, but that would allow a larger current than necessary to flow before the fuse melted. It is safest to choose a fuse rating just above the correct current.

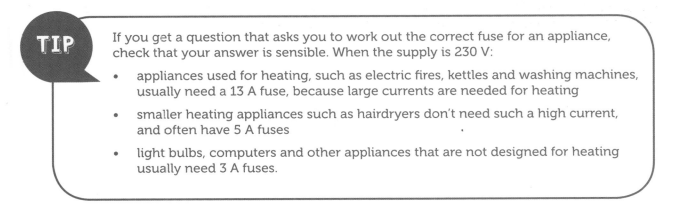

TIP

If you get a question that asks you to work out the correct fuse for an appliance, check that your answer is sensible. When the supply is 230 V:

- appliances used for heating, such as electric fires, kettles and washing machines, usually need a 13 A fuse, because large currents are needed for heating

- smaller heating appliances such as hairdryers don't need such a high current, and often have 5 A fuses

- light bulbs, computers and other appliances that are not designed for heating usually need 3 A fuses.

Direct and alternating current

Cells and batteries produce **direct current**. In a direct current, the electrons always flow in the same direction along the wires.

Electricity produced by generators is **alternating current**. In an alternating current, the electrons flow first one way and then the other. This is because of the way the electricity is generated (see Section 6C).

The mains voltage in many parts of the world has a frequency of 50 Hz, which means that it changes direction 100 times each second. In some places, such as the Caribbean islands, the mains frequency is 60 Hz.

Fig. 2a.06 shows how an oscilloscope can be used to show how the voltage (and so also the current) changes with time for different sources of electricity.

* In Fig. 2a.06(a) the voltage supplied by a cell or battery does not change with time.
* In Fig. 2a.06(b) the voltage of an alternating supply changes with a frequency of 50 Hz. Each complete cycle is $\frac{1}{50}$ second.

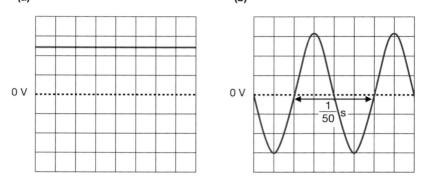

(a)   **(b)**

Fig. 2a.06: Oscilloscope traces:
(a) Constant voltage
(b) Alternating voltage

You should now be able to:

★ give the names and symbols of the units for current, power, voltage and energy (see page 42)

★ describe some hazards of using mains electricity and how to avoid them (see page 43)

★ name some conducting and some insulating materials (see page 43)

★ describe how insulation is used in domestic appliances and what double insulated means (see page 43)

★ explain how earthing, fuses and circuit breakers are used (see page 44)

★ describe some ways in which electrical heating is used (see page 45)

★ explain that a current in a resistor transfers energy (see page 45)

★ recall and use the formula relating power, current and voltage, including rearranging it if necessary (see page 46)

★ use the formula relating energy, current, voltage and time, including rearranging it if necessary (see page 46)

★ explain the difference between alternating current and direct current (see above).

Review questions

1. A 1200 W hairdryer is connected to a 110 V mains supply. What current does it use?

2. A motor runs off a 12 V supply and draws a current of 0.5 A. How long does it take to transfer 90 J of energy?

3. Which of these fuses should be fitted in the plug for a 1 kW hairdryer that runs on mains electricity at 230 V: 3 A, 5 A or 13 A?

Practice questions

1. The picture shows a modern energy-efficient 11 W light bulb.

(a) The bulb is fitted to a table lamp that is plugged into the 230 V mains supply. What current does it use? **(2)**

(b) How much energy is transferred when the bulb is switched on for 1 hour? **(3)**

(c) Which fuse should be used in the plug for the table lamp? Choose from 3 A, 5 A and 13 A, and explain your choice. **(2)**

(d) Why is a fuse used in the plugs for appliances with metal casings? **(4)**

(e) A different table lamp does not have an earth wire. Explain how this lamp can still be safe. **(2)**

B Energy and potential difference in circuits

You will be expected to:

- ★ explain the differences between series and parallel circuits, and how they are used
- ★ describe how current changes when the voltage or the number of components in a circuit changes
- **CAM** ★ work out the combined resistance of resistors in series and in parallel
- ★ recall the way current splits up and recombines in parallel circuits and use this
- ★ recall how the potential difference is divided between components in series and parallel circuits and use this
- ★ describe how current changes with voltage for wires, resistors, filament lamps and diodes
- **CAM** ★ describe how a diode can be used as a rectifier
- ★ describe how to investigate the change in current when the voltage or components are changed
- **CAM** ★ describe how resistance changes with the length and diameter of a wire, and carry out calculations using these ideas
- ★ describe how the resistance of LDRs and thermistors changes, and how these components can be used
- **CAM** ★ describe how capacitors work and how they can be used in time delay circuits
- ★ describe how relays and transistors work
- ★ recall and use the formula relating voltage, current and resistance, including rearranging it if necessary
- ★ recall that current is a rate of flow of charge, and use the formula for this
- ★ describe the difference between electron flow and conventional current
- ★ recall and use the formula relating charge, current and time, including rearranging it if necessary
- ★ recall and use the fact that voltage is the energy transferred per coulomb of charge
- **CAM** ★ explain the terms e.m.f., digital and analogue
- ★ describe how various logic gates work and know their symbols
- ★ understand simple circuits involving logic gates.

Units

You will use the following quantities and units in this section. You should already know most of these from your previous studies.

Quantity	Symbol	Unit
charge	Q	coulombs (C)
current	I	amperes (or amps) (A)
resistance	R	ohms (Ω)
time	t	seconds (s)
voltage/potential difference	V	volts (V)

You also need to recall and use the standard symbols for electrical components. A full list of these is given in Appendix 2 (on page 242).

Series and parallel circuits

In a **series circuit**, all the components are arranged within one loop of wire.

- The same current flows through all the components.
- One switch controls all the components at once.
- If a bulb breaks in a series circuit, it will create a gap in the circuit and none of the other components will work.

In a **parallel circuit**, there are different 'branches' for the current to flow through.

- A switch in the main part of the circuit will control all the components at once.
- Each component can also be controlled individually by putting a switch in its branch.
- If a set of bulbs is wired in parallel in a circuit and one bulb breaks, the other bulbs will continue to work.

In the parallel circuit in Fig. 2b.01:

- switch S1 can be used to turn all the bulbs off at once
- bulb A can be switched on by closing S1 and S2
- bulb B can be switched on by closing S1 and S3
- if bulb A breaks, bulb B will still work as long as the correct switches are closed.

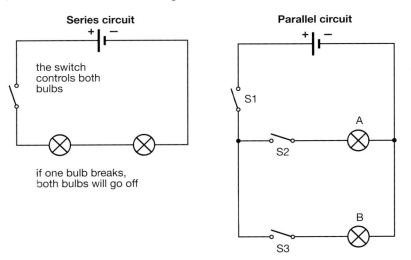

Fig. 2b.01: Series and parallel circuits

Lights in houses are all on parallel circuits. This allows each light to be switched on and off individually. It also means that if one bulb breaks, the whole house does not go dark, and extra bulbs can be added without making the other ones dimmer.

Current in series circuits

The current in a circuit depends on the voltage and also on the number and type of components in the circuit. If the components stay the same, increasing the voltage increases the current.

Every component has a resistance. For a certain voltage, the higher the resistance in a circuit, the smaller the current. Adding bulbs or other components to a circuit increases the total resistance of the circuit, so the current gets smaller.

CAM

Resistors in series and in parallel

The resistance of a component is measured in ohms (Ω). If a circuit contains two or more resistors, the total resistance of the circuit can be calculated.

For a series circuit, the total resistance of the circuit is the sum of the resistances of all the components (Fig. 2b.02).

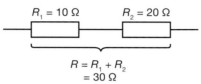

$$R = R_1 + R_2$$
$$= 30\ \Omega$$

Fig. 2b.02: The combined resistance of resistors in series is the sum of the individual resistances

In a parallel circuit, the current from the cell has more than one path that it can follow. The combined resistance of two resistors in parallel is less than the resistance of either of the resistors individually.

The effective resistance of two resistors in parallel is given by this formula:

$$\frac{1}{R} = \frac{1}{R_1} + \frac{1}{R_2}$$

Worked example

What is the total resistance of the circuit in Fig. 2b.03 (on page 53)?

Answer

$$\frac{1}{R} = \frac{1}{10} + \frac{1}{20}$$
$$= 0.1 + 0.05$$
$$\frac{1}{R} = 0.15$$
$$R = 6.67\ \Omega$$

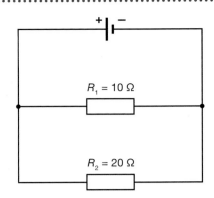

Fig. 2b.03: Resistors in a parallel circuit

Current and voltage in series and in parallel circuits

In a series circuit, the voltage from the cell is divided between the components in the circuit. The higher the resistance of a component, the greater its share of the total voltage (Fig. 2b.04).

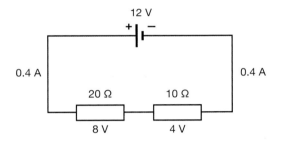
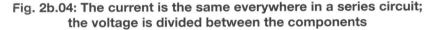

Fig. 2b.04: The current is the same everywhere in a series circuit; the voltage is divided between the components

Current flowing in a parallel circuit splits up when it reaches a junction, and recombines when the wires meet at another junction. The current from the cell is the sum of the currents in the branches. The voltage is the same in each branch of a parallel circuit.

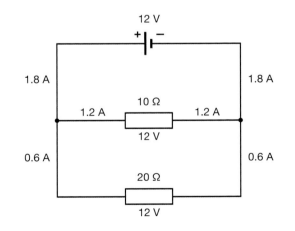

Fig. 2b.05: Current and voltage in a parallel circuit

Potential divider circuits

A potential divider circuit uses resistors or other components to provide different voltages to other components in the circuit, as shown in Fig. 2b.06. If one of the resistances can be changed, the circuit is a **variable potential divider** (also called a **potentiometer**).

The power supply for the circuit is connected to the two ends of the variable resistor. A parallel circuit is set up between one end of the resistor and the sliding contact. Moving the sliding contact changes the voltage in the parallel circuit shown in Fig. 2b.06(b).

- Fig. 2b.06(a) shows a fixed potential divider circuit. The voltmeters (and the rest of the circuits connected across the resistors) are in parallel with the resistors. The potential difference across the 20 Ω resistor is two-thirds of the total voltage, because 20 Ω is two-thirds of the total resistance of the circuit.

- Fig. 2b.06(b) shows a variable potential divider circuit. The lowest voltages in the external circuit are with the slider to the left.

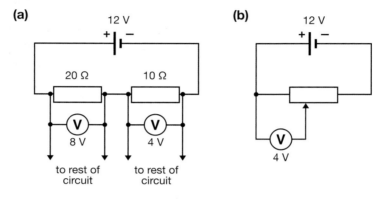

Fig. 2b.06: (a) A fixed potential divider circuit
(b) A variable potential divider circuit

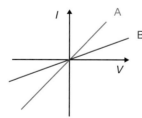

TIP   Many electronic circuits depend on the idea of a potential divider, so make sure you understand how this circuit works.

Current in different components

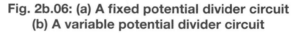

Fig. 2b.07 shows how the current changes in two different wires or resistors when the voltage is changed. The straight lines on this graph show that the resistance of each component is constant. The negative values of current and voltage show what happens if the current flows in the opposite direction.

Fig. 2b.07: Current–voltage graphs for two resistors: resistor A has a lower resistance than resistor B, because there is more current for a given voltage

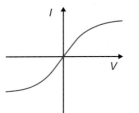

Metal filament lamps get hot when they are switched on. If the voltage is increased, the current increases and the filament gets hotter. High temperatures increase the resistance of the wire in the filament lamp, and so the current–voltage graph for a filament lamp is curved (Fig. 2b.08).

Fig 2b.08: A filament lamp, and a current–voltage graph for the lamp

(a)

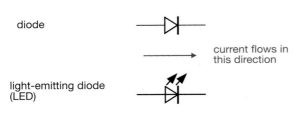

A **diode** is an electronic component that allows current to flow in one direction but not the other. As shown in Fig. 2b.10 a diode needs a small **forward** voltage of about 0.6 V before it will conduct. If the voltage is applied in the opposite direction (a '**reverse**' voltage) it will not conduct. A **light emitting diode** (or **LED**) is a diode that lights up when current flows through it. Light bulbs and LEDs in a circuit can show whether or not current is flowing.

(b)

diode

current flows in this direction

light-emitting diode (LED)

TIP Current flows through a diode or LED in the direction of the 'arrowhead' part of the symbol.

Fig. 2b.09: (a) Diodes and LEDs (b) Their symbols

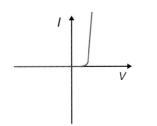

Fig. 2b.10: Current–voltage graph for a diode

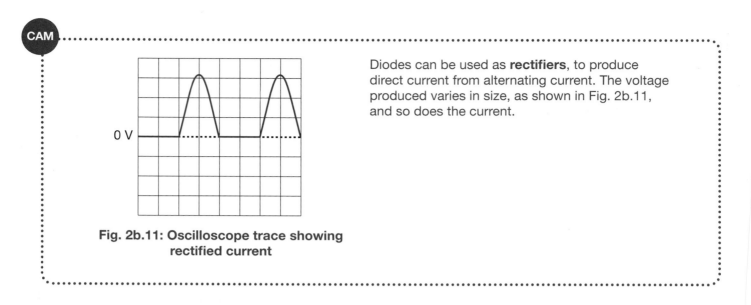

Diodes can be used as **rectifiers**, to produce direct current from alternating current. The voltage produced varies in size, as shown in Fig. 2b.11, and so does the current.

Fig. 2b.11: Oscilloscope trace showing rectified current

Investigating current and voltage

The way current changes with voltage for different components can be investigated using a circuit such as the one shown in Fig. 2b.12. School power supplies have different voltage settings, but they do not always produce exactly the voltage set on them. This is why you need to measure the voltage across the component as well as the current.

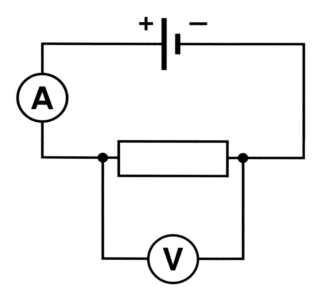

Fig. 2b.12: Circuit used to investigate how current changes with voltage: this circuit contains a resistor, but other components can be put in its place

Resistance of wires

The resistance of a wire depends on the type of metal that it is made from, but it also depends on the diameter and the length of the wire.

*   The resistance is proportional to the length of the wire. If the length of wire is doubled, the resistance doubles.
*   The resistance is inversely proportional to the cross-sectional area of the wire (i.e. proportional to $\frac{1}{A}$). If the area doubles, the resistance is halved, and if the area is halved, the resistance doubles.

Worked example

Wires A, B, and C are all made of the same metal. Wire A is 0.5 m long and has a resistance of 3 Ω. Wire B has the same cross-sectional area as wire A and is 2 m long. Wire C is 0.5 m long and has twice the cross-sectional area of wire A. What are the resistances of wires B and C?

Answer

Wire B is 4 times as long as wire A, so its resistance is 4 times that of A.

Resistance of wire B = 4 × 3 Ω

$\qquad\qquad\qquad$ = 12 Ω

Wire C has twice the cross-sectional area of wire A, so its resistance is half that of A.

Resistance of wire C = $\dfrac{3\ \Omega}{2}$

$\qquad\qquad\qquad$ = 1.5 Ω

LDRs and thermistors

The resistance of a **light-dependent resistor (LDR)** changes as the amount of light falling on it changes. The brighter the light, the lower the resistance.

The resistance of a **thermistor** changes when the temperature changes. The higher the temperature, the lower the resistance.

LDRs and thermistors can be used in circuits to control things automatically. For example:

*   an LDR can be used in a circuit to switch on outside lights when it gets dark
*   a thermistor could be used to control the heating system in a house.

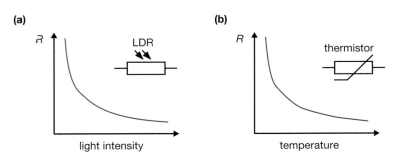

Fig. 2b.13: (a) How the resistance of an LDR changes with light intensity
(b) How the resistance of a thermistor changes with temperature

Capacitors and time delay circuits

A **capacitor** consists of two metal plates separated by an insulator (which can be air or a solid material). The capacitor can act as an energy store (Fig. 2b.14).

Electrons pushed around the circuit by the cell collect on one plate and their negative charges repel electrons from the opposite plate. Current flows and the bulb lights up until the capacitor is fully charged.

If the cell is removed, the capacitor discharges. The bulb lights up until the capacitor has discharged.

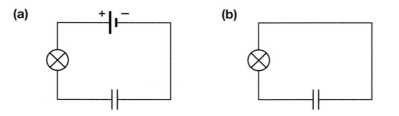

Fig. 2b.14: (a) Current will flow in this circuit until the capacitor is fully charged (b) The capacitor in this circuit will discharge; the bulb will light up until the capacitor has discharged

Capacitors can be used in **time delay circuits**. An example of a time delay circuit is shown in Fig. 2b.15. The LED in this circuit lights up several seconds after the circuit has been switched on. The length of the delay depends on the size of the resistors and the capacitor.

- When the circuit in Fig. 2b.15 is first switched on, the capacitor is charging so its resistance is low. The voltage between A and B is therefore much higher than the voltage between B and C.
- As the LED and its resistor are in parallel with the capacitor, there is the same, low voltage across them and the LED does not light up.
- As the capacitor charges up, its resistance increases.
- When it is fully charged, its resistance is much greater than resistor X, and so the voltage across BC is much greater than the voltage across AB.
- The voltage across BC is also the voltage across the LED, which is now high enough to make it light up.

Fig. 2b.15: A time delay circuit

> **TIP**
> In diagrams showing electronic circuits, it is usual to show the positive voltage supply at the top of the diagram and zero volts at the bottom, rather than drawing a complete circuit with a battery or power supply. This makes it easier to show how the components are arranged, and easier to build a circuit based on the diagram.

Relays and transistors

A **relay** is an electrically operated switch. A small current in a relay can be used to switch on a circuit carrying a much higher current. There is more on relays in Section 6B.

Most relays consist of a coil of wire wrapped around an iron core. When a current flows through the coil it becomes magnetised, and this attracts the switch and closes the circuit. When the current in the coil stops, a spring returns the switch to its open position. Fig. 2b.17 (on page 60) shows an example of a relay being used in a control circuit for an electric heater.

A **transistor** is another form of electrically operated switch. It has no moving parts. It can be used to amplify a current, as shown in Fig. 2b.16.

- In this circuit, a sensor is used to detect when the temperature is above a certain level. However, the sensor does not produce a large enough current to light the bulb.
- When there is no current at all from the sensor, the transistor is 'off' and no current can flow between X and Y.
- When there is a small current from the sensor going to the base of the transistor, it allows current to flow from X to Y and the bulb lights. The current has been amplified because a small current from the sensor allows a larger current to flow through the bulb.

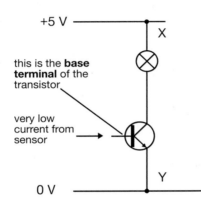

Fig. 2b.16: Part of a circuit with a transistor

Using electronics for control

Thermistors and light-dependent resistors can be used as **input transducers** in control circuits. This means that they convert variations in a physical quantity (temperature or light, in this example) into an electrical signal.

Fig. 2b.17 shows a thermistor being used to switch on a heater when the temperature is below a certain value. The variable resistor allows the temperature at which the heater comes on to be adjusted.

- When the temperature is high, the resistance of the thermistor is low and so there is a low voltage across the relay coil.
- When the temperature falls, the resistance of the thermistor is high compared to that of the variable resistor. There is a higher voltage across the relay coil, so a higher current flows. The relay coil is magnetised and closes the switch in the heater circuit.

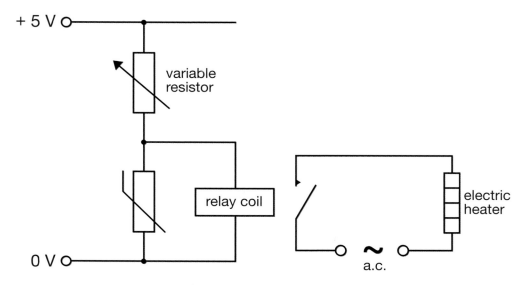

Fig. 2b.17: Using a thermistor to switch a heater on

If the variable resistor and thermistor are swapped over, there will be a high voltage across the relay coil when the temperature is high. This circuit could be used to operate an alarm if the temperature goes above a certain value.

If an LDR is used in place of the thermistor, the circuit can be used to switch lights on when it gets dark, or to close shades in a greenhouse if the light gets too bright.

Resistance, current and voltage

The resistance, voltage and current for a resistor or wire at a constant temperature are related by this formula:

voltage = current × resistance

$$V = I \times R$$

TIP — You need to remember this formula as it will not be given to you in the exam. You also need to be able to rearrange it.

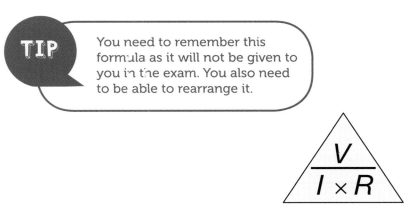

Fig. 2b.18: Formula triangle for resistance, current and voltage

You can demonstrate that $V = I \times R$ is correct for a standard resistor using the apparatus shown in Fig. 2b.12 (on page 56).

Worked example

A 5 Ω resistor is connected to a 2 V cell. What current flows in the circuit?

Answer

$$\text{current} = \frac{\text{voltage}}{\text{resistance}}$$

$$= \frac{2\ V}{5\ \Omega}$$

$$= 0.4\ A$$

Current and charge

A current is the rate of flow of **charge**. In the case of a current flowing in a wire, the moving charges are electrons. Electrons have a negative charge.

The size of the current depends on how much charge flows past a point in a particular time. Charge is measured in **coulombs (C)**.

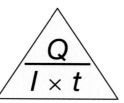

TIP In a circuit with a cell, electrons flow out of the negative terminal of the cell and back into the positive one. This is the opposite direction to **conventional current**, which is assumed to flow from + to −.

When you are thinking about which way current flows through a diode, or about electromagnetic induction (Section 6), you need to use the direction of conventional current.

Current, charge and time are related by this formula:

$$\text{charge} = \text{current} \times \text{time}$$

$$Q = I \times t$$

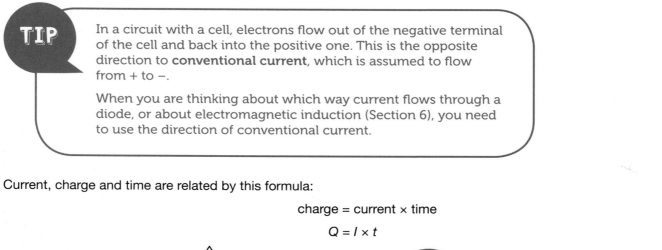

TIP You need to remember this formula as it will not be given to you in the exam. You also need to be able to rearrange it.

Fig. 2b.19: Formula triangle for charge, current and time

Voltage

The **voltage** is a measure of the amount of energy transferred by a current. Voltage is sometimes called **potential difference**.

A voltmeter effectively measures the energy of the current entering a component and compares it with the energy leaving it. The voltage (or potential difference) is the number of joules of energy that have been transferred in the component for every coulomb of charge flowing through it.

1 volt is 1 joule transferred for each coulomb of charge.

The **e.m.f.** of a source of electricity is the amount of energy supplied by a cell or other power supply. It is the energy supplied per coulomb of charge, and is also measured in volts. E.m.f stands for 'electromotive force', but it is not really a force.

Logic gates

Logic gates are used in **digital** circuits. Digital signals are either on or off (shown by 1 and 0). **Analogue** signals are signals that can vary continuously.

Logic gates are circuits that contain transistors and other components, designed to give an output that depends on the input or inputs.

Logic gates and their symbols

Name	NOT	AND	OR	NAND	NOR
Symbol					
Description	• also called an inverter • reverses the input signal	• the output is on (1) if both inputs are on (if input 1 AND input 2 are on)	• the output is on (1) if either of the inputs is on (if input 1 OR input 2 is on)	• a NOT gate combined with an AND gate • the output is off (0) if both the inputs are on	• a NOT gate combined with an OR gate • the output is off (0) if either of the inputs is on

A **truth table** shows the output of the gate for each different combination of inputs.

The tables below show the truth tables for the logic gates you need to know about.

NOT gate

Input	Output
0	1
1	0

AND gate

Input A	Input B	Output
0	0	0
0	1	0
1	0	0
1	1	1

OR gate

Input A	Input B	Output
0	0	0
0	1	1
1	0	1
1	1	1

NAND gate

Input A	Input B	Output
0	0	1
0	1	1
1	0	1
1	1	0

NOR gate

Input A	Input B	Output
0	0	1
0	1	0
1	0	0
1	1	0

TIP You don't need to memorise the outputs of NAND and NOR gates. Just write down what the AND and OR gates would do, and then reverse the outputs.

Logic gates can be used in combinations to carry out different control tasks.

Example

A homeowner has a conservatory that gets very hot in warm weather. During the day, she keeps it cool by closing the blinds. If it is night-time or it is cloudy, then a fan is used to blow air through the conservatory instead. She uses the combination of logic gates shown in Fig. 2b.20.

During the day, the light sensor output is 1, and the NOT gate changes this to 0. So one input of AND gate B is always 0 during the day, and the output is 0. This means that the fan cannot be switched on during the day, no matter what the output of the temperature sensors.

If the temperature is high, the output of the temperature sensor is 1. If it is daylight, then both inputs to logic gate A are 1. The output is 1 and the motor is operated to close the blinds. If it is night-time, the input to A from the light sensor is 0 and so the motor will not operate.

If the temperature is high at night, the two inputs to AND gate B will be 1 and the fan will be switched on.

Further logic gates and controls would be needed to make the blinds open again if the temperature dropped during the day.

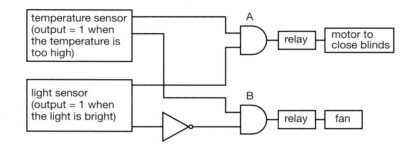

Fig. 2b.20: Logic circuit for controlling the temperature in a conservatory

You should now be able to:

★ describe the differences between series and parallel circuits (see page 51)

★ explain why parallel circuits are used for wiring lights in a house (see page 51)

★ describe how current changes when the voltage or the number of components in a circuit changes (see page 52)

CAM ★ know how to work out the combined resistance of resistors in series and in parallel (see page 52)

★ describe how current splits up and recombines at junctions in parallel circuits (see page 53)

★ describe how the potential difference is divided between components in series and parallel circuits (see page 54)

★ sketch graphs to show how current changes with voltage for wires, resistors, filament lamps and diodes (see page 54)

CAM ★ describe how a diode can be used as a rectifier (see page 56)

★ describe how to investigate the change in current when the voltage or components are changed (see page 56)

CAM ★ describe how resistance changes with the length and diameter of a wire (see page 57)

★ describe how the resistance of LDRs and thermistors changes, and how these components can be used (see page 57)

CAM ★ describe how capacitors work and how they can be used in time delay circuits (see page 58)

★ describe how relays and transistors work (see page 59)

★ recall and rearrange the formula relating voltage, current and resistance (see page 61)

★ describe current as a rate of flow of charge, and use the formula for this (see page 62)

★ describe the difference between electron flow and conventional current (see page 62)

★ recall and rearrange the formula relating charge, current and time (see page 62)

★ describe how voltage is the energy transferred per coulomb of charge (see page 62)

CAM ★ explain the terms e.m.f., digital and analogue (see page 63)

★ describe how various logic gates work and know their symbols (see page 63)

★ understand simple circuits involving logic gates (see page 64).

Review questions

1. How will the following changes affect the current in a series circuit:

   (a) replacing a 20 Ω resistor with a 10 Ω resistor

   (b) increasing the voltage of the electricity supply?

CAM 2. There are two 10 Ω resistors in a circuit. What is the total resistance of the circuit if it is:

   (a) a series circuit

   (b) a parallel circuit?

3. A series circuit has two bulbs in it and uses a 12 V supply. What is the potential difference across each bulb?

4. What is a transducer?

5. A 12 V supply causes a 2 A current to flow in a circuit. What is the total resistance of the circuit?

6. 100 C of charge flows through a bulb in 0.5 seconds. What is the current?

CAM 7. A control circuit is used to switch on a fan if the temperature gets too hot during the day.

   (a) Which two transducers would be in the circuit?

   (b) What kind of logic gate would be used?

Practice questions

1. The circuit shows the wiring for lights in a combined kitchen and dining room. There are two lights in the kitchen, which are both controlled by one switch, and just one light in the dining end of the room.

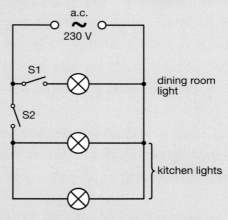

   (a) Explain why the two kitchen lights are in parallel with the dining room light. **(1)**

   (b) Explain why the kitchen lights are in parallel with each other, even though they are both controlled by the same switch. **(1)**

   (c) Explain what would happen if an extra bulb was placed in series with the existing dining room light. **(2)**

CAM 2. Each bulb in the circuit in question 1 has a resistance of 2500 Ω.

   (a) What is the total resistance of the bulbs in the kitchen? **(2)**

   (b) What current flows through switch S1 when it is closed? **(3)**

3. This graph shows how the current changes when the voltage across a filament lamp changes.

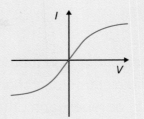

(a) Describe and explain three key features of this graph. **(6)**

(b) How does the graph show that the resistance increases when the *temperature* of the filament increases? Include a formula to help you to explain. **(4)**

CAM 4. The diagram below shows a time delay circuit. When it is switched on, the LED does not light immediately but comes on after a short time.

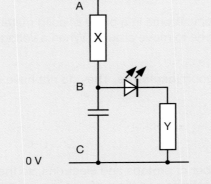

(a) Explain why decreasing the resistance of resistor X would decrease the time delay of the circuit. **(2)**

(b) Explain why the LED will remain lit for a few seconds after the circuit is switched off. **(3)**

5. Car headlamps run from a 12 V battery. One bulb draws a current of 4 A.

(a) What is the resistance of the bulb? **(2)**

(b) How much charge flows through the bulb when it is switched on for 10 seconds? **(3)**

(c) How much energy is transferred in this time? Use your answer to part (b) to help you to work this out. **(2)**

(d) Check your answer to part (c) using a different formula. **(2)**

CAM 6. A greenhouse is fitted with an automatic watering system. It is not good for plants to be watered in bright sunshine, so an LDR is included in the circuit. The LDR part of the circuit is on when the light is bright, and the water sensor part is on when the plants are moist.

(a) Sketch a logic circuit (like Fig. 2b.20) to show a circuit that could be used to control the relay that pumps water into the greenhouse. The system should also show a warning light when the soil is dry. **(4)**

(b) Explain your choices. **(4)**

C Electric charge

You will be expected to:

★ recall some materials which are electrical conductors and insulators
★ explain how positive and negative electrostatic charges are produced
★ recall how electric charges attract or repel each other
CAM ★ describe some simple electric field patterns
★ explain some electrostatic phenomena in terms of the movement of electrons
★ describe some dangers of electrostatic charges
★ describe some uses of electrostatic charges.

Conductors and insulators

Metals are good **conductors** of electricity. Atoms in a piece of solid metal are held together in a fixed pattern, with some of the electrons from the atoms free to move around. When a voltage is applied to the metal, these electrons all move in the same direction.

Materials such as wood or plastic are good **insulators**. They do not have electrons that can move around freely, so they do not conduct electricity.

Electrostatic charges

Materials normally have the same number of protons and electrons, so the positive charges on the protons balance the negative charges on the electrons and the material has no overall charge.

If you rub two pieces of material together, electrons can be transferred from one to the other. If a material gains electrons by being rubbed, it then has an overall negative charge. The material it was rubbed with has lost these electrons, so now it has more protons than electrons, and has an overall positive charge.

If a metal is rubbed, electrons can be transferred, but because the metal conducts electricity the charges can spread out throughout the metal and we do not usually notice their effects. When an insulating material is rubbed, the charges that have been transferred cannot move around. These are called **electrostatic** charges.

TIP It is always the electrons that are transferred when a charge is produced by rubbing. This is because the electrons are on the outside of the atoms. Protons are held in the nucleus of each atom, and cannot be transferred when materials are rubbed together.

Two positively charged objects will repel each other, and two negatively charged objects will also repel each other. Negative and positive charges attract each other.

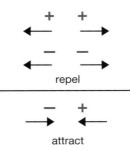

Fig. 2c.01: Unlike charges attract and like charges repel

Electric fields

An **electric field** is a region where an electric charge experiences a force.

The direction of the field is the direction of the force it would exert on a positive charge, so the field goes from positive to negative. You need to learn the shapes of the fields around point charges and between parallel plates, as shown in Fig. 2c.02.

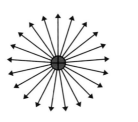

Fig. 2c.02: Electric fields around point charges and between parallel plates

Electrostatic effects

(a)   **(b)**

Fig. 2c.03: Some electrostatic effects

The machine in Fig. 2c.03(a) is a Van de Graaff generator, which produces a charge of static electricity on the dome. Some of the charge is transferred to the girl touching the dome, and spreads out. All the strands of her hair have the same charge, so they all repel each other and her hair stands on end.

Fig. 2c.03(b) shows a comb picking up bits of paper. This happens because the electrostatic charge on the comb **induces** an opposite charge in the pieces of paper and the two charges attract each other.

If a negatively charged balloon is held close to a wall, the negative charge on the balloon repels electrons in the wall (Fig. 2c.04). The surface of the wall now has a positive charge, so it attracts the negatively charged balloon.

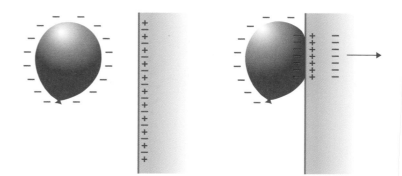

Fig. 2c.04: Charging by induction can be used to make a balloon stick to a wall

Dangers of electrostatic charges

If enough charges build up on an object they can jump to an uncharged object as a spark. You may have felt this yourself when touching a car door handle, or after walking across a nylon carpet and opening a door.

When you feel a spark it may be slightly painful, but if the same thing happens while an aircraft is being refuelled it can cause an explosion.

* To prevent this, the aircraft and the tanker refuelling it are joined with a wire before the fuel tanks are opened, and the wire is also attached to earth.
* The wire allows any static charges that have built up on the aircraft to travel along it.
* If the tanker and the aircraft have both been earthed they will have no static charges that could make a spark.

Using electrostatic charges

Electrostatic charges are used in paint sprayers, inkjet printers and photocopiers.

Inkjet printer

An inkjet printer gives each drop of ink an electrostatic charge. The drops of ink travel between two metal plates before they hit the paper. The two metal plates are given electric charges, and the size and sign of the charges can change many times each second. The changing charges on the plates direct the ink drops to different places on the paper.

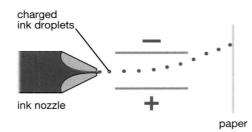

Fig. 2c.05: In an inkjet printer, charged ink drops are directed using electrostatic charges

Photocopier

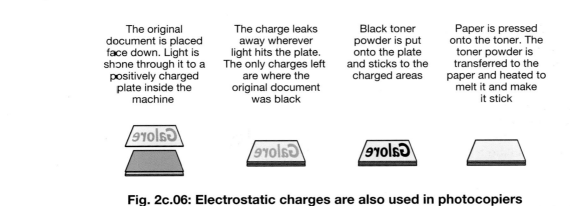

| The original document is placed face down. Light is shone through it to a positively charged plate inside the machine | The charge leaks away wherever light hits the plate. The only charges left are where the original document was black | Black toner powder is put onto the plate and sticks to the charged areas | Paper is pressed onto the toner. The toner powder is transferred to the paper and heated to melt it and make it stick |

Fig. 2c.06: Electrostatic charges are also used in photocopiers

You should now be able to:

★ name some materials which are electrical conductors (see page 68)

★ name some materials which are electrical insulators (see page 68)

★ explain why a plastic comb gets a positive electrostatic charge after being used to comb hair (see page 68)

★ describe how electric charges attract or repel each other (see page 69)

CAM ★ describe some simple electric field patterns (see page 69)

★ describe and explain what happens when two balloons are rubbed with a cloth, and then suspended close to each other (see page 70)

★ describe some dangers of electrostatic charges (see page 70)

★ describe how electrostatic charges are used in inkjet printers and photocopiers (see above).

Review questions

1. Draw the shape of the electric field around a small object with a negative charge.

2. If you rub a plastic comb hard and then hold it next to a thin stream of water from a tap, you can make the stream of water bend. Explain why this happens.

3. Why are fuel tankers connected to the aeroplane and to earth before refuelling starts?

4. In a photocopier, why is the pattern on the document being copied converted into a pattern of charges on a plate?

Practice questions

1. A Van de Graaff generator produces static electricity on a metal dome. A student stands with one hand on the dome and a small plastic container of puffed rice breakfast cereal in the other. After a minute the puffed rice starts jumping out of the container.

   (a) Explain why the student should not be touching any other metal objects during this demonstration. **(1)**

   (b) Why does the puffed rice fly out of the container? **(3)**

   (c) In a second demonstration, another student blows bubbles towards the Van de Graaff generator. The student is not touching the generator. Explain why the bubbles are attracted towards the machine. **(3)**

2. Fuel for petrol mowers, strimmers and other tools is often stored in small, metal fuel containers. The following safety advice is given by a local fire service:

   > • Put the metal container on the ground when filling it up at a filling station – do not fill the container while it is still in the back of a vehicle.
   >
   > • Ensure that the metal nozzle of the fuel pump stays in contact with the metal container while it is being filled.

   (a) Why is static electricity a danger at filling stations? **(2)**

   (b) Explain why the ground in filling stations is usually made of a material that has a relatively low electrical resistance. **(3)**

   (c) Explain the precautions in the advice given above. **(4)**

3. Insecticides are sprayed onto crops. In some sprayers, the nozzle that dispenses the spray is given an electric charge. Suggest why this is done. **(3)**

Section Three

3 Waves

A Properties of waves

You will be expected to:

★ describe longitudinal and transverse waves
★ explain the meanings of amplitude, frequency, wavelength and time period
★ recall and use the fact that waves transfer energy without transferring matter
★ recall and use the formula for speed, frequency and wavelength of a wave
★ use the relationship between frequency and time period
★ describe diffraction and how it depends on the size of a gap.

Units

You will use the following quantities and units in this section.

Quantity	Symbol	Unit
frequency	f	hertz (Hz)
speed	v	metres/second (m/s)
time	t	seconds (s)
time period	T	seconds (s)
wavelength	λ	metres (m)

Longitudinal and transverse waves

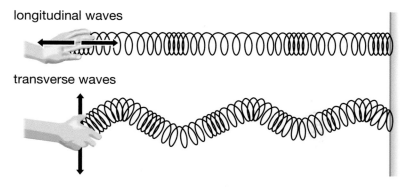

longitudinal waves

transverse waves

Fig. 3a.01: 'Slinky' springs can be used to demonstrate different types of wave

Waves transfer energy without transferring matter. Waves can be transverse or longitudinal (see Fig. 3a.01).

You need to remember whether some common waves are transverse or longitudinal.

Wave	Transverse	Longitudinal
water	✓	
spring	✓	✓
rope	✓	
light	✓	
sound		✓

Describing waves

Waves can be described by using their speed, amplitude, frequency, wavelength and period.

- The **speed** is the distance a wave travels in one second.
- The **amplitude** is the maximum displacement of the wave from the zero position.
- The **frequency** is the number of waves that pass a point each second.
- The **wavelength** is the distance between two peaks or two troughs of a wave.
- The **time period** is the time for one complete cycle of a wave.

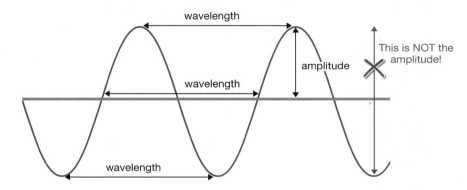

Fig. 3a.02: Wavelength and amplitude for a transverse wave: the wavelength can be measured in any of the places shown

TIP

Make sure you get the amplitude correct. It is the distance from the undisturbed position, *not* the distance from top to bottom of the wave.

Wave equations

Wave speed, frequency and wavelength are linked by the following formula:

$$\text{wave speed} = \text{frequency} \times \text{wavelength}$$

$$v = f \times \lambda$$

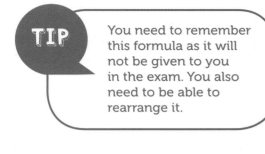

TIP You need to remember this formula as it will not be given to you in the exam. You also need to be able to rearrange it.

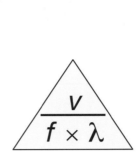

Fig. 3a.03: Formula triangle for wave speed, frequency and wavelength

The frequency and time period of a wave are linked by this formula:

$$\text{frequency} = \frac{1}{\text{time period}}$$

$$f = \frac{1}{T}$$

This can also be written as:

$$T = \frac{1}{f}$$

Worked example

A light wave travels at 3×10^8 m/s. It has a wavelength of 0.5 cm. What are its frequency and time period?

Answer

The wavelength must be in metres.

$$0.5 \text{ cm} = 0.005 \text{ m}$$

$$\text{frequency} = \frac{\text{speed}}{\text{wavelength}}$$

$$= \frac{3 \times 10^8 \text{ m/s}}{0.005 \text{ m}}$$

$$= 6 \times 10^{10} \text{ Hz}$$

$$\text{time period} = \frac{1}{\text{frequency}}$$

$$= \frac{1}{6 \times 10^{10} \text{ Hz}}$$

$$= 1.67 \times 10^{-11} \text{ s}$$

Water waves

Waves in water can be used to help us to understand what happens when waves are reflected, refracted or diffracted. (There is more on reflection and refraction in Section 3C, starting on page 84.)

You may have seen waves demonstrated using a **ripple tank**. This is a shallow tank of water with a transparent base and a 'dipper' that bobs in and out of the water to make waves. Light shines through the water and the base, and the waves make shadows. The ripple tank can be used to investigate reflection by putting a barrier in the tank.

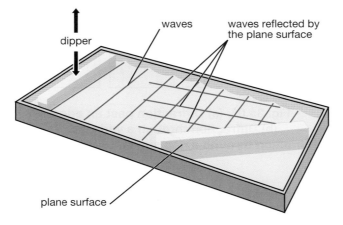

Fig. 3a.04: A ripple tank

Refraction is the change in direction of waves caused by a change in the speed they travel. The speed of water waves depends on the depth of the water. We can use this idea to demonstrate refraction in a ripple tank by placing a block in the water to change its depth.

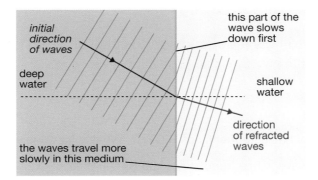

Fig. 3a.05: A ripple tank shows why waves change direction when they slow down

Water waves can also be used to demonstrate diffraction, by placing a barrier with a gap in a ripple tank (see page 78).

Diffraction

Waves passing an edge or passing through a gap spread out (Fig. 3a.06). This effect is called **diffraction**.

The amount of diffraction depends on:

- the wavelength
- the size of the gap.

There is most diffraction when the size of the gap is similar to the wavelength of the waves.

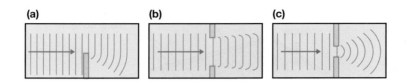

**Fig. 3a.06: (a) Diffraction past an edge
(b) Diffraction through a wide gap
(c) Diffraction through a gap similar to the wavelength**

You should now be able to:

★ give the names and symbols for the units of frequency, wavelength and time period (see page 74)
★ give some examples of transverse and longitudinal waves (see page 75)
★ sketch a transverse wave and mark its amplitude and wavelength (see page 75)
★ recall and use the formula for speed, frequency and wavelength of a wave (see page 76)
★ describe diffraction and how it depends on the size of a gap (see above).

Review questions

1. A wave travels at 330 m/s and has a frequency of 1000 Hz. What is its wavelength?

2. What is the time period of a wave with a frequency of 50 Hz?

3. You are standing outside an open doorway. You cannot see into the room but you can hear people inside talking. Use the idea of diffraction to explain why. (*Hint*: think about the wavelengths of the different waves and the size of the doorway.)

Practice questions

1. A tsunami is a huge wave in the ocean, caused by an earthquake or underwater landslide. Tsunami warning systems include a set of pressure sensors on the sea bed that can detect tsunamis by pressure changes as the wave passes over them. The sensors are linked to buoys on the surface that transmit the data to a control centre.

    After an earthquake in the Pacific Ocean, one pressure sensor detected a tsunami wave at 05.10, and then another, smaller wave at 05.30. A buoy 3000 km away detects the first wave at 09.20 (15 000 seconds after the sensor).

    (a) What kind of wave is a wave on the surface of water? **(1)**

    (b) How fast is the tsunami travelling? **(2)**

    (c) What is the period of the tsunami? Give your answer in seconds. **(2)**

    (d) What is its frequency? **(2)**

    (e) What is its wavelength? **(2)**

2. The diagram shows a small harbour on the coast. Waves usually approach the harbour from the direction shown. 'Choppy' waves caused by local winds have wavelengths from about 10 to 20 m, whereas long period 'swells' caused by distant storms can have wavelengths up to 200 m or more.

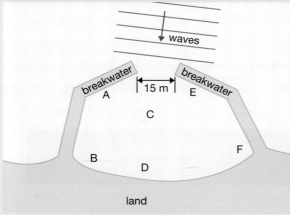

    (a) Explain where the best places are to anchor a boat in a swell. **(3)**

    (b) Explain where the best places are for anchoring if the sea is choppy. **(3)**

B The electromagnetic spectrum

You will be expected to:

★ recall the different parts of the electromagnetic spectrum, in order
★ recall that all electromagnetic waves travel at the same speed in a vacuum
★ recall some of the uses of electromagnetic waves
★ recall some of the harmful effects of excessive exposure to some electromagnetic waves and some protective measures that can be taken.

The electromagnetic spectrum

Light is part of the **electromagnetic spectrum**. The waves in the electromagnetic spectrum all travel at the same speed in **free space** (a vacuum, sometimes referred to as *in vacuo*). This speed is approximately 3×10^8 m/s.

Although the electromagnetic spectrum is continuous, for convenience, it is divided up into different parts, as shown in Fig. 3b.01. Radio waves have the longest wavelengths and lowest frequencies, and gamma rays have the shortest wavelengths and highest frequencies.

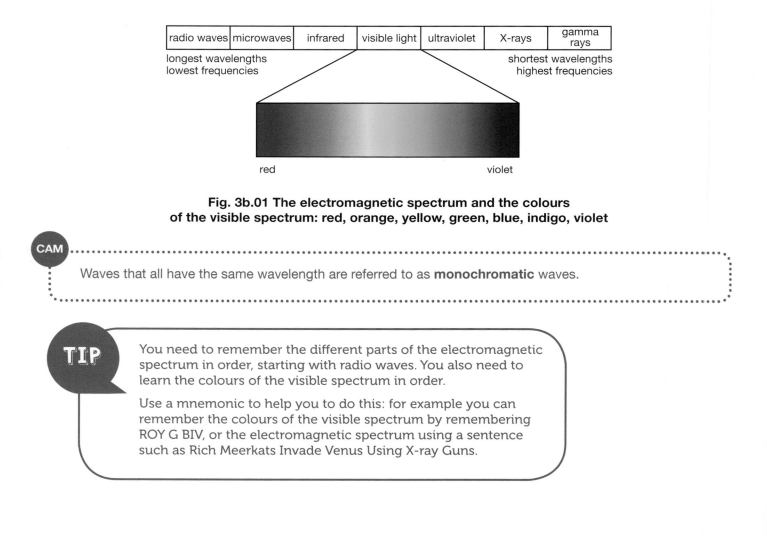

**Fig. 3b.01 The electromagnetic spectrum and the colours
of the visible spectrum: red, orange, yellow, green, blue, indigo, violet**

CAM

Waves that all have the same wavelength are referred to as **monochromatic** waves.

TIP

You need to remember the different parts of the electromagnetic spectrum in order, starting with radio waves. You also need to learn the colours of the visible spectrum in order.

Use a mnemonic to help you to do this: for example you can remember the colours of the visible spectrum by remembering ROY G BIV, or the electromagnetic spectrum using a sentence such as Rich Meerkats Invade Venus Using X-ray Guns.

Using the electromagnetic spectrum

Radio waves can travel through walls and other solid objects, and can be diffracted around hills. They are used:

- for broadcasting radio and television programmes
- in communications.

Microwaves travel in straighter lines than radio waves because their wavelength is shorter. They are absorbed by water, and transfer energy to water as heat. They are used for:

- cooking (in microwave ovens)
- transmitting TV programmes via satellite
- transmitting telephone calls.

Infrared radiation is emitted by all bodies – warmer bodies emit more radiation than cooler ones. Infrared radiation:

- is emitted by heaters such as fires, grills and toasters
- is emitted by remote control handsets and detected by TVs, DVD players, etc.
- is detected by sensors in burglar alarms
- is detected by night vision equipment and equipment used by emergency services to locate people trapped in damaged buildings.

Visible light:

- is used to send information along optical fibres
- is detected by cameras.

Ultraviolet light can be absorbed by certain chemicals that then give out visible light. Fluorescent lamps work in this way. Ultraviolet light is also used:

- to detect security markings on banknotes and other items
- in sunbeds to cause tanning of skin.

X-rays can pass through some materials but are absorbed by others. They are used:

- to make images of the inside of the body
- to check the internal structure of metal objects
- in airports and ports to check luggage for dangerous items.

Gamma rays can kill cells, including bacteria. They are used for:

- sterilising surgical instruments for operations
- irradiating food (to kill bacteria so the food lasts longer)
- detecting and treating cancer.

Dangers of the electromagnetic spectrum

The human body is exposed to most of the electromagnetic spectrum. Too much radiation can harm the body. The harm done depends on how much radiation we absorb. The more radiation that is absorbed, the greater is the damage.

The amount of damage done also depends on the wavelengths to which we are exposed.

* Microwaves can cause internal heating of body tissues ('cooking' the body), which is why microwave ovens have shielding and door locks that stop users opening the door while microwaves are being produced.
* Infrared radiation can cause skin burns.
* Ultraviolet radiation can damage surface skin cells and may lead to skin cancer. It also damages the eyes and can eventually lead to blindness. Harm can be avoided by covering up in the sunniest part of the day, using sun cream and wearing sunglasses.
* Gamma rays damage cells by causing mutations in the DNA, which can lead to cancer. Workers using sources of gamma rays follow strict safety rules to keep their dose as low as possible.

You should now be able to:

* ★ recall the parts of the electromagnetic spectrum in order, starting with the ones with the longest wavelengths (see page 80)
* ★ describe some uses of each type of wave (see page 81)
* ★ describe some of the harmful effects of microwaves, infrared radiation, ultraviolet radiation and gamma rays (see above)
* ★ describe some simple precautions to avoid harm from electromagnetic waves (see above).

Review questions

1. Which type (or types) of wave is used for the following purposes?

   (a) TV remote control handsets

   (b) mobile phone signals

   (c) finding people in collapsed buildings

   (d) taking photographs

   (e) detecting security markings

   (f) sterilising surgical instruments

   (g) making images of suspected broken bones

   (h) cooking

   (i) sending information along optical fibres.

2. Why do people trapped in collapsed buildings show up on infrared cameras?

3. Which electromagnetic waves can:

   (a) damage the eyes

   (b) lead to cancer

   (c) be used to cook food?

Practice questions

1. William Herschel is best known for his discovery of Uranus in 1781, but he also experimented with light. In 1800 he used a prism to create a spectrum from sunlight, and placed thermometers in different parts of the spectrum. He found that the temperature reached by the thermometer increased as he moved his thermometers from the violet to the red end of the spectrum. He also found that the thermometer showed a temperature increase when it was placed beyond the red end of the visible spectrum.

   (a) Name the colours of the visible spectrum in order, starting with red. **(1)**

   (b) Which part of the electromagnetic spectrum was Herschel investigating when his thermometer was placed beyond the red part of the visible spectrum? **(1)**

   (c) (i) What would he be investigating if his thermometer were placed beyond the other end of the spectrum? **(1)**

   (ii) Name two effects of this radiation on the human body. **(2)**

   (d) Name one feature that all the waves in the electromagnetic spectrum have in common. **(1)**

   (e) Which waves in the electromagnetic spectrum have:

   (i) the shortest wavelengths **(1)**

   (ii) the lowest frequencies? **(1)**

   (f) Name two uses for:

   (i) X-rays **(2)**

   (ii) gamma rays. **(2)**

C Light and sound

You will be expected to:

* ★ know that light waves are transverse waves that can be reflected, refracted and diffracted
* ★ recall and use the law of reflection
* ★ draw ray diagrams to show how an image is formed in a mirror
* ★ describe experiments to investigate the refraction of light
* ★ recall and use the formula for the refractive index
* **CAM** ★ draw diagrams to show how images are formed by lenses
* ★ explain how light can be dispersed using a prism
* ★ explain what the critical angle is and recall and use the formula for the critical angle
* ★ describe how total internal reflection is used in optical fibres
* ★ describe the advantages of using digital signals rather than analogue signals
* ★ show that sound waves are longitudinal waves which can be reflected, refracted and diffracted
* ★ understand how to use an oscilloscope to display sound waves
* ★ state the hearing range for humans
* ★ describe how to measure the speed of sound in air
* ★ explain how pitch and loudness are related to frequency and amplitude.

Units

You will use the following quantities and units in this section.

Quantity	Symbol	Unit
angle		degrees (°)
angle of incidence	i	degrees (°)
angle of refraction	r	degrees (°)
refractive index	n	
time period	T	seconds (s)

Light waves and reflection

Light waves are transverse waves that can be reflected, refracted and diffracted. Diffraction is covered on page 78.

TIP Make sure you remember the difference between reflection, refraction and diffraction:

* reflection happens when waves 'bounce off' a barrier
* refraction is when waves change direction because they have changed speed
* diffraction is a spreading out of waves as they pass a barrier.

When waves are reflected by a surface, the **angle of incidence** is equal to the **angle of reflection**. This is called the **law of reflection**.

Fig. 3c.01 shows water waves hitting a barrier, with arrows showing the direction of movement of the waves. This is a model for what happens to light waves. When we draw diagrams to represent light or sound, we usually just draw these arrowed lines.

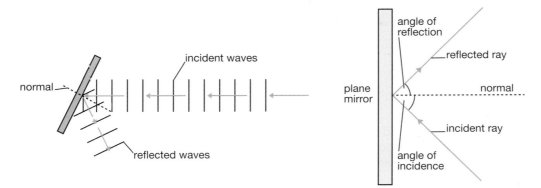

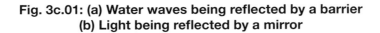

Fig. 3c.01: (a) Water waves being reflected by a barrier
(b) Light being reflected by a mirror

> **TIP**
> Remember that all the angles used in describing the behaviour of light are measured from the **normal**, not from the surface of a mirror or glass block.

Images can be **real** or **virtual**. The image formed on a screen by a projector is a **real** image. You can touch the screen where the image is formed.

The image in a mirror is a **virtual** image. The image appears to be *behind* the mirror. The ray diagram in Fig. 3c.02 shows how this virtual image is formed. You need to be able to draw a ray diagram like this.

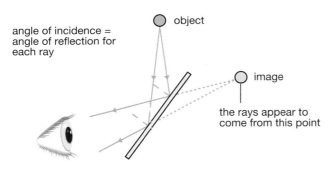

Fig. 3c.02: Ray diagram for a virtual image in a mirror

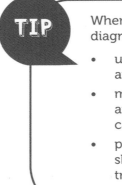

> **TIP**
> When you are drawing ray diagrams:
> - use a sharp pencil, a ruler and a protractor
> - measure and draw all the angles as carefully as you can
> - put arrows on the rays to show which way the light is travelling.

An image in a mirror is:

- virtual
- upright
- laterally inverted (left and right are swapped over)
- the same distance behind the mirror as the object is in front of the mirror
- the same size as the object.

Investigating waves

Reflection and refraction can be investigated using ray boxes and **ray tracing**. A ray box is used to send a thin beam of light towards a mirror or a glass block. The position of the ray is marked using small crosses, and then the crosses are joined with a ruler to show where the ray of light went.

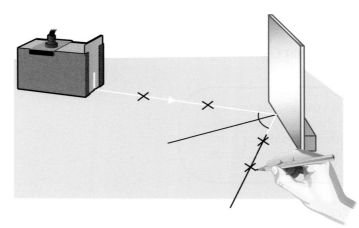

Fig. 3c.03: Ray tracing is used to investigate reflection and refraction

Refraction

Light waves travel at different speeds in different materials. The most common materials you will be asked about are glass and water.

Light travels at almost the same speed in air as it does in a vacuum, but it travels more slowly in water and even more slowly in glass. The change in speed causes **refraction** if the waves meet the **interface** between the two materials at an angle to the normal. The waves do not change direction if they meet the interface along the normal (at 90° to the surface).

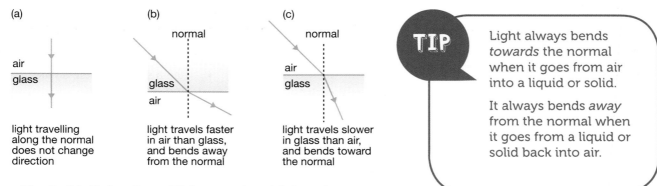

(a) light travelling along the normal does not change direction

(b) light travels faster in air than glass, and bends away from the normal

(c) light travels slower in glass than air, and bends toward the normal

TIP

Light always bends *towards* the normal when it goes from air into a liquid or solid.

It always bends *away* from the normal when it goes from a liquid or solid back into air.

Fig. 3c.04: Refraction of light at a glass/air interface

The amount of refraction is measured by comparing the angle of incidence with the angle of refraction (the angle between the refracted ray and the normal). These angles are used to calculate the **refractive index** (n):

$$n = \frac{\sin i}{\sin r}$$

Note that n is a ratio, so it has no units.

The refractive index depends on the two materials involved, and which way the light is travelling. The refractive index for light going from air to glass is not the same as the refractive index for light going from glass to air.

Fig. 3c.05: Formula triangle for the refractive index

TIP You need to remember this formula as it will not be given to you in the exam. You also need to be able to rearrange it.

TIP If you are calculating an angle of incidence or refraction, don't forget to turn the sine back into an angle.

Worked examples

Example 1

In Fig. 3c.04(c) the angle of incidence is 45° and the angle of refraction is 28°. What is the refractive index for light going from air to glass?

Answer

$$n = \frac{\sin i}{\sin r}$$
$$= \frac{\sin 45°}{\sin 28°}$$
$$= \frac{0.707}{0.469}$$
$$= 1.5$$

Example 2

The refractive index for light passing from water to air is 0.75. The angle of refraction is 60°. What is the angle of incidence?

Answer

$$\sin i = n \times \sin r$$
$$= 0.75 \times \sin 60°$$
$$= 0.75 \times 0.866$$
$$= 0.6495$$
$$r = 40.5°$$

The refractive index can also be calculated using the speed of light in the different materials. For example, the speed of light in air is approximately 3×10^8 m/s, and its speed in glass is approximately 2×10^8 m/s.

Example

$$\text{refractive index for air to glass} = \frac{\text{speed of light in air}}{\text{speed of light in glass}}$$

$$= \frac{3 \times 10^8 \text{ m/s}}{2 \times 10^8 \text{ m/s}}$$

$$= 1.5$$

Finding the refractive index by experiment

You can find the refractive index of a glass block by:

- placing the glass block on a piece of paper and drawing around it
- aiming a light ray at the block and marking where the ray goes across the paper by putting at least two crosses on each line (Fig. 3c.06(a))
- removing the block and joining up the crosses using a ruler
- joining the points where the light entered and left the block (Fig. 3c.06(b))
- measuring the angles of incidence and refraction
- calculating the value of n.

You should repeat the process at least three times and take a mean value of n, discarding any results that are very different from the others (see page 201).

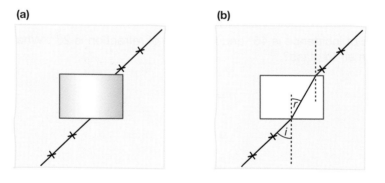

(a)   (b)

Fig. 3c.06: Finding the refractive index of a glass block

Refraction by prisms

Different wavelengths of light change speed by different amounts when they enter a new material. This means that the different colours that make up white light are refracted by different amounts.

You can see this best using a **prism**. Violet light is refracted more than red light as the light enters the prism, and again when it leaves. White light is split up into the colours of the **spectrum**. This effect is called **dispersion**.

Monochromatic light (light of a single colour, which has a narrow range of wavelengths) does not form a spectrum.

red

violet

Fig. 3c.07: A triangular prism can disperse visible light

Lenses

Lenses are made of glass or transparent plastic, and their shape allows light passing through them to form images.

Converging lenses bring parallel rays of light to a focus. The point at which parallel rays converge to a point after passing through a converging lens is called the **principal focus**, and the distance between the centre of the lens and the principal focus is the **focal length** of the lens.

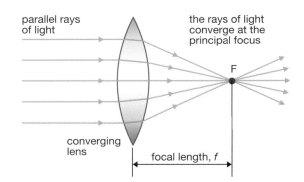

parallel rays
of light

the rays of light
converge at the
principal focus

F

converging
lens

focal length, *f*

Fig. 3c.08: A converging lens

TIP

Diverging lenses are thinner in the middle than at the edges and make light spread out. You are not required to know about this type of lens for the CIE specification.

The type of image formed by a lens depends on where the object is. You can draw ray diagrams to work out where an image will be. There are three important rays you need to draw on a ray diagram (Fig. 3c.09):

- a ray that passes straight through the centre of the lens without changing direction
- a ray from the object parallel to the axis of the lens, which is bent to pass through the principal focus on the far side of the lens
- a ray from the object that passes through the principal focus on the same side of the lens, which emerges parallel to the lens axis (this ray cannot be drawn if the object is less than one focal length from the lens).

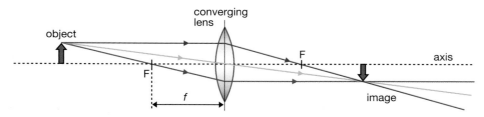
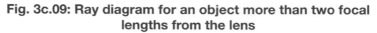

Fig. 3c.09: Ray diagram for an object more than two focal lengths from the lens

If the object is further than one focal length from the lens the image will be:

- real
- inverted
- on the opposite side of the lens to the object.

The size and exact position of the image will depend on the distance of the object from the lens.

If the object is closer than one focal length to the lens, the lens is being used as a magnifying glass (Fig. 3c.10). The image formed by a magnifying glass is:

- virtual
- upright
- magnified.

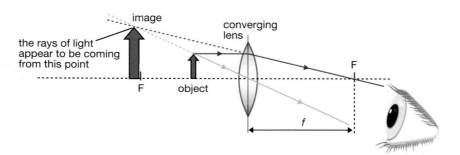
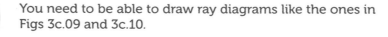

Fig. 3c.10: Ray diagram for a magnifying glass

TIP

You need to be able to draw ray diagrams like the ones in Figs 3c.09 and 3c.10.

Do your drawings carefully and accurately. If you can draw three lines (as in Fig. 3c.09), your three rays of light should cross at one point if you have made an accurate drawing.

Physics A* Study Guide

Total internal reflection

Light bends away from the normal when it leaves glass or water and enters air. As the angle of incidence increases, the angle of refraction also increases.

Eventually the angle of refraction becomes 90° and the ray of light lies along the edge of the glass or along the surface of the water. This angle of incidence is called the **critical angle**, and is given the symbol **c**.

If the angle of incidence is increased further, the light is reflected from the inside of the block. This is called **total internal reflection**.

The critical angle for a material depends on the refractive index for the material:

$$\sin c = \frac{1}{n}$$

Worked example

In this example, the refractive index for light passing from water to air is 1.33. What is the critical angle for water?

Answer

$$\sin c = \frac{1}{1.33}$$

$$= 0.75$$

$$c = 48.6°$$

(If you don't round your answer to the first part of this calculation you should get 48.8°. It is important to show all your working, so an examiner knows why you may have got a slightly different answer.)

 TIP Total internal reflection only happens when light is passing *out* of water or glass and into air.

TIP Even though the critical angle applies to light leaving water or glass and going into air, the value for *n* used in the calculation is always the value for light going *into* the material. Don't worry about this, as the exam question will give you the number you need to use.

A semicircular block is used to find the critical angle for a material (Fig. 3c.11).

A ray of light pointed through the block at the centre of the flat side passes through the curved side at 90°, so its direction does not change. The ray of light is moved around until the light is just reflected internally. The angle of incidence at which this happens is the critical angle.

Fig. 3c.11: Light being totally internally reflected inside a semicircular glass block

Using total internal reflection

Light can be totally internally reflected inside a prism. Prisms are used inside binoculars and inside cameras instead of mirrors. Mirrors sometimes produce multiple images if the light is reflected from the front and the back surfaces of the glass, but this does not happen with prisms.

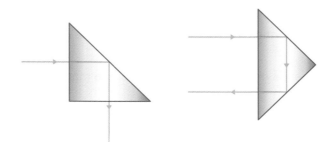

Fig. 3c.12: Total internal reflection can also happen inside prisms

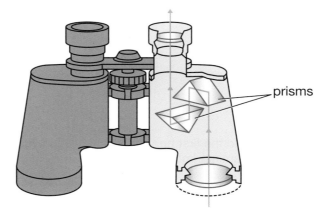

Fig. 3c.13: Binoculars use two prisms for each eye; without the prisms the binoculars would be longer and the image would be upside down

Total internal reflection is also used to send light along **optical fibres**. These are made of thin strands of glass, with the outer layer of glass having a higher refractive index than the centre of the fibre. Light sent into the fibre at one end is repeatedly reflected along the fibre.

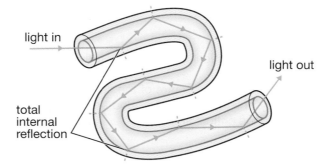

Fig. 3c.14: Light is totally internally reflected inside optical fibres

Optical fibres are used in **endoscopes**. These are medical instruments used to help doctors to see inside parts of the body such as the stomach or intestines. An endoscope consists of two bundles of optical fibres which are put down the patient's throat. One bundle transmits light, which is reflected from the inside of the patient. The other bundle sends the reflected light back to an eyepiece or camera.

Optical fibres are also used in communications, because:

- signals can be sent over longer distances without amplification than with copper wires
- optical fibres do not suffer from interference from nearby power lines.

Analogue and digital signals

An **analogue signal** varies continuously. The sound produced by a TV or radio is an analogue signal. A **digital signal** is a series of pulses, referred to as 'on' and 'off', or 1 and 0. Analogue signals can be converted into digital signals.

Almost all signals need to be amplified before they are used. The amplification can happen inside a TV or radio, or it can be done along the transmission line if signals are being sent long distances.

When analogue signals are amplified, any distortions or interference they have picked up will also be amplified. With digital signals, because the signal is either a 0 or a 1, the amplified signal can be made to look just the same as the original signal.

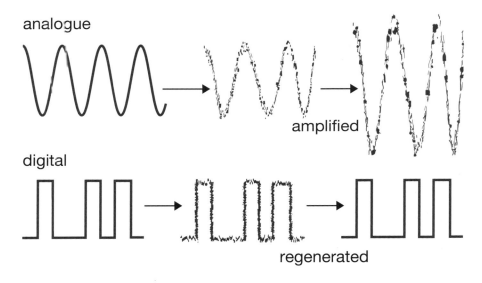

Fig. 3c.15: The effects of amplification on analogue and digital signals

Analogue broadcasts of TV or radio programmes modify the frequency of the waves, so each programme needs a range of frequencies. In the UK, all TV broadcasts are being changed to digital. The 'on or off' digital signal can be sent using just one frequency. This means that digital signals can carry more information than analogue ones and many more TV channels can be broadcast using the same range of frequencies.

Sound waves

Sound waves are longitudinal waves produced by vibrating sources. They need a **medium** to travel through, so they can only travel through solids, liquids or gases. Sound waves cannot travel through a vacuum.

As they are waves, they can be reflected, refracted and diffracted:

* when sound is reflected you hear an echo
* sounds travel further at night because sound is refracted by layers of air in the atmosphere at different temperatures
* you can hear sounds from around a corner because sound is diffracted.

Sound waves consist of a series of compressions and rarefactions (Fig. 3c.16). Air particles are squashed closer together in the compressions, and are further apart in the rarefactions. The wavelength is the distance between one rarefaction and the next, or between one compression and the next.

A microphone and oscilloscope can be used to display a sound wave. The display shows a shape that looks like a transverse wave, but it is important to remember that this is only a *representation* of the sound wave. A sound wave is longitudinal.

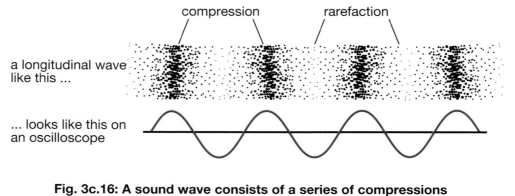

Fig. 3c.16: A sound wave consists of a series of compressions and rarefactions; the dots represent air particles

CAM

Sound travels at approximately 340 m/s in air, but the exact speed varies with the temperature. Sound travels faster in liquids than in gases, and faster still in solids. For example, the speed of sound in water is approximately 1500 m/s and the speed of sound in steel is nearly 6000 m/s.

Loudness and pitch

The **amplitude** of a sound wave is a measure of how far the particles move when the wave passes. The greater the amplitude, the further each particle moves during each wave and the louder the sound.

The **pitch** of a sound is how high or low it sounds. The pitch depends on the frequency of the sound waves. Sound waves with high frequencies have high pitches. Humans can hear sounds in the range 20 Hz to 20 000 Hz.

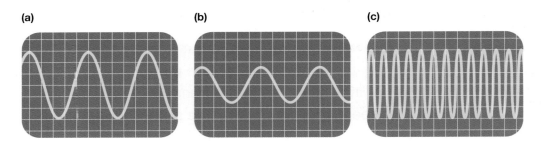

(a)   (b)   (c)

**Fig. 3c.17: (a) Two complete waves shown on an oscilloscope
(b) This sound has the same pitch as sound (a), but is quieter
(c) This sound has a higher pitch than sound (a) and has the same loudness**

An oscilloscope can be used to work out the frequency of a sound wave, as shown in the example below.

> **Worked example**
>
> On the traces shown in Fig. 3c.17(a) and (b), each square represents a time of 0.001 seconds. Each complete wave takes 0.005 seconds, so the time period of the wave is 0.005 seconds. Calculate the frequency.
>
> *Answer*
>
> $$\text{frequency} = \frac{1}{\text{time period}}$$
> $$= \frac{1}{0.005 \text{ s}}$$
> $$= 200 \text{ Hz}$$

Measuring the speed of sound

You can measure the speed of sound in air using simple laboratory equipment.

One way is to measure a distance from a large wall then clap or bang two pieces of wood together and listen for the echo.

- You can work out the speed of sound from the distance and the time between the clap and the echo, remembering that the sound has travelled to the wall *and back*.
- It is difficult to measure this time accurately with a stopwatch, so it is better to clap in time with the echo, so your second clap happens when you hear the first echo.
- Once you have built up a rhythm, get a friend to time ten claps, and then work out the speed from the total distance and the total time.

You should now be able to:

★ explain the terms reflection, refraction and diffraction (see page 84)

★ state the law of reflection (see page 85)

★ draw ray diagrams to show how an image is formed in a mirror (see page 85)

★ explain the difference between virtual and real images (see page 85)

★ describe experiments to investigate the refraction of light (see pages 86 and 88)

★ recall which way light bends when it goes into and out of glass or water (see page 86)

★ recall and use the formula for the refractive index (see page 87)

CAM ★ draw diagrams to show how images are formed by lenses (see page 89)

★ explain how light can be dispersed using a prism (see page 89)

★ explain what the critical angle is (see page 91)

★ recall and use the formula for the critical angle (see page 91)

★ describe how total internal reflection is used in optical fibres (see page 92)

★ describe the difference between analogue and digital signals (see page 93)

★ describe the advantages of using digital signals rather than analogue signals (see page 93)

★ describe how sound waves are longitudinal waves which can be reflected, refracted and diffracted (see page 94)

★ understand how to use an oscilloscope to display sound waves (see page 94)

★ state the hearing range for humans (see page 95)

★ describe how to measure the speed of sound in air (see page 95)

★ explain how pitch and loudness are related to frequency and amplitude (see page 95).

Review questions

1. A light ray is reflected with an angle of reflection of 30°. What was the angle of incidence?

2. The refractive index for light going from air into water is 1.33. What is the angle of refraction if the angle of incidence is 30°?

3. The refractive index for diamond is approximately 2.4. What is the critical angle for diamond?

4. How does increasing the frequency of a sound wave change (a) its pitch, (b) its loudness?

CAM 5. Describe three rays that can be used in ray diagrams to work out the kind of image produced by a lens.

6. Write down two properties of an image formed by an object more than one focal length from a converging lens.

7. (a) Where does the object have to be when a converging lens is being used as a magnifying glass?

   (b) Describe the image formed by a magnifying glass.

Practice questions

1. The diagram shows a candle in front of a mirror. Two light rays from the candle flame are shown.

    (a) Copy the diagram and finish drawing the two rays. Show the position of the image of the candle flame. **(3)**

    (b) Mark one angle of incidence and one angle of reflection on your diagram. **(1)**

    (c) State the law of reflection that you have used to help you to draw the rays. **(1)**

CAM 2. A converging lens has a focal length of 3 cm.

    (a) Draw a ray diagram to show the image formed by an object at 5 cm from the lens. You may find it helpful to draw your diagram on graph paper. **(3)**

    (b) Describe the image formed. **(3)**

3. The refractive index for light passing from air to glass is 1.5.
The refractive index for light passing from glass into air is 0.66.
The critical angle for light passing from glass into water is approximately 62°.

    (a) If light leaving glass and passing into air has an angle of incidence of 20°, what is the angle of refraction? **(3)**

    (b) What is the critical angle for light passing from glass to air? (Use $n = 1.5$) **(3)**

Some cars have windscreen wipers that switch on automatically when there is water on the windscreen. A beam of infrared light is sent through the windscreen glass with an angle of incidence at the inside surface of the glass of 45°.

outside

windscreen glass

inside of car

detector

    (c) When the windscreen is dry, the infrared will be passing from glass into air. Explain what will happen to the beam of infrared light if its angle of incidence is 45°. **(2)**

    (d) Explain what will happen to the beam of infrared light if there is a raindrop on the windscreen. **(2)**

    (e) Explain how the automatic sensor determines when to switch on the wipers. **(2)**

4. Broadcast TV programmes in the UK are gradually being converted from analogue to digital broadcasts.

    (a) What is the difference between analogue and digital signals? **(2)**

    (b) Describe two advantages of using digital signals instead of analogue signals. **(2)**

5. A student is using a signal generator and oscilloscope to measure the speed of sound in air. She sets the signal generator to a frequency of 1200 Hz and connects it to a loudspeaker.

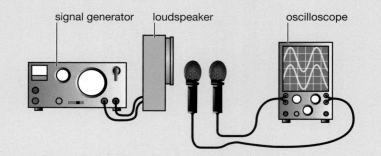

She connects two microphones to the oscilloscope. One microphone is placed fairly close to the loudspeaker, and it produces a trace on the oscilloscope screen. She moves the other microphone away from the first one until the oscilloscope produces a trace similar to the one shown below. The two microphones are now one wavelength apart.

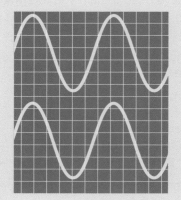

(a) What kind of wave is a sound wave? **(1)**

(b) What feature of the trace above shows the loudness of the sound? Include a sketch with your answer. **(2)**

(c) What does the distance between two peaks of a wave *on the oscilloscope* trace represent? Explain your answer. **(2)**

The table shows the student's results.

Test	Distance between microphones (m)
1	0.283
2	0.286
3	0.276
4	0.279
5	0.285

(d) What is the mean distance? **(1)**

(e) Suggest why the student repeated the measurement several times. **(1)**

(f) What is the speed of sound measured by the student? **(3)**

Section Four

4 Energy resources and energy transfer

A Energy transfer

You will be expected to:

★ describe energy transfers between different forms of energy
★ understand that energy is conserved
★ calculate the efficiency of energy transfers
★ represent energy transfers using Sankey diagrams
★ explain how energy is transferred by conduction, convection and radiation
★ give some everyday examples of convection
★ describe how insulation is used to reduce energy transfers.

Units

You will use the following quantity and unit in this section.

Quantity	Symbol	Unit
energy	E	joules (J)

Forms of energy

Everything around us happens because energy is transferred.

It helps us to think about energy transfers by considering different forms of energy. You should be able to give some examples of energy transfers involving all the forms of energy below.

- **Thermal** (heat) energy is transferred from hotter to cooler objects, for instance when a hot drink cools down. Thermal energy is sometimes called **internal** energy.
- **Light** energy is emitted by light bulbs, fires and by some chemical reactions.
- **Electrical** energy is transferred when an electric current flows through a conductor.
- **Sound** energy is transferred by vibrations in solids, liquids or gases.
- **Kinetic** (movement) energy is energy stored in moving objects.
- **Chemical** energy is stored in the bonds between atoms in molecules and can be transferred in chemical reactions.
- **Nuclear** energy is stored within atoms, and can be transferred by radiation or nuclear reactions (see Section 7).
- **Gravitational potential** energy is energy stored in objects in raised positions.
- **Elastic potential** energy is also called strain energy – it is stored in stretched, compressed or twisted objects.

You can show energy changes using a flow diagram. For example, when an arrow is fired from a bow:

chemical energy $\longrightarrow$ elastic potential energy $\longrightarrow$ kinetic energy
(in muscles)   (in stretched bow)   (in moving arrow)

Efficiency

Whenever anything happens, energy is transferred in some way. Energy is **conserved**, which means that the *total* energy before and after the transfer remains the same. Energy cannot be created or destroyed, just transferred from one place to another or from one form to another.

In any energy transfer, some of the transferred energy is in a useful form, but some is wasted because it is in a form that we do not want. For example, an electric fire transfers electrical energy to useful heat energy and wasted light energy (Fig. 4a.01). The heat energy is useful, because we are using the fire to heat the room. The light energy is wasted energy in this case, because we are not using the electric fire to light the room.

the light energy is not useful because we are not using the fire to light the room

the heat energy is useful because we are using the fire to heat the room

Fig. 4a.01: Energy transfers in an electric fire

The efficiency of an energy transfer is a measure of how much of the energy is wasted. For most energy transfers, the wasted form of energy is heat energy, but a small amount of energy is often wasted as sound energy as well.

$$\text{efficiency} = \frac{\text{useful energy output}}{\text{total energy input}}$$

Efficiency is a ratio, so it has no units.

CAM

This is sometimes expressed as a percentage:

$$\text{efficiency} = \frac{\text{useful energy output}}{\text{energy input}} \times 100\%$$

Worked example

A filament light bulb transfers 100 J of energy every second. It produces 91 J of heat energy. What is its efficiency?

Answer

First you need to work out the amount of useful energy transferred each second.

$$\text{useful energy output} = \text{total energy input} - \text{wasted energy output}$$
$$= 100\ \text{J} - 91\ \text{J}$$
$$= 9\ \text{J}$$
$$\text{efficiency} = \frac{\text{useful energy output}}{\text{total energy input}}$$
$$= \frac{9\ \text{J}}{100\ \text{J}}$$
$$= 0.09$$

TIP Efficiency can never be greater than 1 (or 100%), as this would mean that more energy was being output than was being put in. If you work out an efficiency greater than 1, go back and check your work because you have got something wrong!

Measuring efficiency

You can measure efficiency in the laboratory by carefully measuring the energy put into a system and the energy that comes out.

One way of doing this is to use an electric motor to lift a load. The energy input can be calculated from the current and voltage used by the motor and the time it is running (see page 46). The useful energy transferred to the load can be calculated from its mass and the distance it is raised (see page 111).

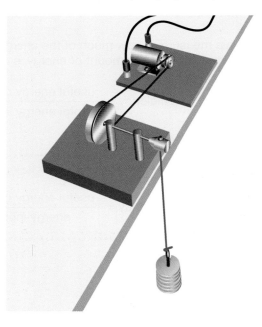

Fig. 4a.02: Using an electric motor to lift a load

Sankey diagrams

A **Sankey diagram** is a way of representing energy transfers that makes it easy to see the relative amounts of energy transferred to different forms.

The key point to remember about Sankey diagrams is that the widths of the arrows are proportional to the amounts of energy they represent.

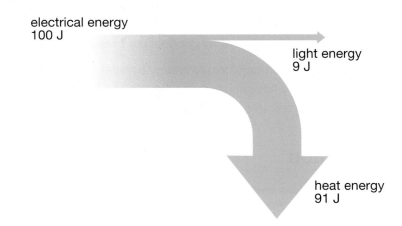

electrical energy
100 J

light energy
9 J

heat energy
91 J

Fig. 4a.03: A Sankey diagram for the light bulb in the example on page 102

Worked example

An energy-saving light bulb transfers 20 J of energy every second, and produces 11 J of light energy. Draw a Sankey diagram to represent the energy transfers in the bulb.

Answer

The left-hand end of the Sankey diagram represents 20 J.

- This Sankey diagram will be easier to draw if you make the arrow on the input side 20 mm wide.
- The output arrow representing light will then be 11 mm wide.
- The part representing the wasted heat energy will be 9 mm wide.
- Label the arrows with the amounts of energy.

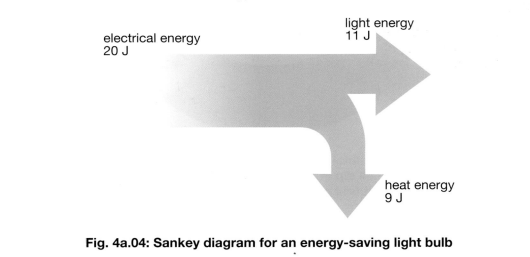

electrical energy
20 J

light energy
11 J

heat energy
9 J

Fig. 4a.04: Sankey diagram for an energy-saving light bulb

TIP

If you are asked to draw a Sankey diagram, make it easier for yourself by choosing a suitable scale. In the example, 1 mm represents 1 J.

If you are asked to represent larger amounts of energy, you could use 1 cm to represent 1000 J.

Heat transfers

Heat energy can be transferred by conduction, convection and radiation.

- **Conduction** is the main form of energy transfer in solid objects.
- **Convection** only happens in fluids (liquids and gases).
- **Radiation** can transfer energy across empty space, and through transparent materials.

Conduction

Conduction transfers energy through solids. In solids the particles are in fixed positions and cannot move around. The particles *can* vibrate about their fixed positions. The hotter the object, the greater these vibrations.

Vibrating particles in one part of an object can pass on the vibrations to nearby particles, and this is how energy is transferred by conduction. Metals are good conductors of heat because the free electrons can also transfer energy.

CAM

You can investigate conduction by heating one end of different materials and measuring the temperature a fixed distance away from the source of heat.

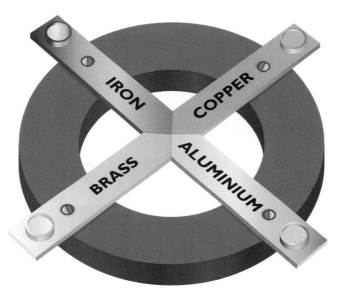

Fig. 4a.05: The metal strips are heated where they meet in the middle; the small dip at the end of each strip contains wax; the wax will melt first on the strip that is the best conductor

Convection

Energy is transferred through **fluids** by convection. Liquids and gases are both classed as fluids because the particles in them can move around.

Convection occurs when part of a fluid is warmed more than the surrounding fluid. Warming part of a fluid makes the particles move around faster and so they take up more space.

The same mass of particles now occupies a larger volume, so its density is less (see Section 5A) and it rises. Cooler fluid moving in to take its place sets up a **convection current**.

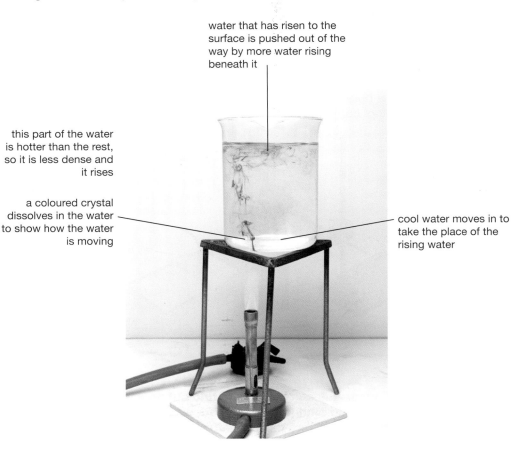

water that has risen to the surface is pushed out of the way by more water rising beneath it

this part of the water is hotter than the rest, so it is less dense and it rises

a coloured crystal dissolves in the water to show how the water is moving

cool water moves in to take the place of the rising water

Fig. 4a.06: A lab demonstration of a convection current

> **TIP**
> You have probably heard the expression 'hot air rises'. This is not quite right – hot air (or any fluid) only rises if it is warmer than the fluid surrounding it. It is the *difference* in density that leads to convection.

Convection currents can also form around cold objects, such as around an ice lolly. In this case, heat transferred from the air to the lolly cools the air, which becomes more dense and sinks.

> **TIP**
> Never say that 'cold' is transferred from the lolly to the air. Heat is a form of energy; cold is the *absence* of heat. Only heat energy can be transferred, so if a cold object 'cools' the air around it, what is really happening is that heat energy is being transferred *to* the cold object from its warmer surroundings. As they have lost energy, the surroundings become cooler.

Radiation

Heat energy can be transferred by **infrared radiation**. This is part of the electromagnetic spectrum.

All objects emit infrared radiation, and the hotter the object the more radiation it emits. Heat is transferred by radiation through air and through a vacuum, and can also be transferred through transparent materials.

When radiation falls on an object, it may be reflected or absorbed. The surface of an object determines how well it absorbs and emits infrared radiation.

- Light, shiny surfaces are poor emitters and absorbers of radiation.
- Dark, dull surfaces are good emitters and absorbers of radiation.

CAM

Fig. 4a.07 shows some apparatus that can be used to investigate how well different colours absorb or emit radiation. Radiation from the bulb in the ray box heats up the two thermometers. When the ray box is switched off, the two thermometers cool at different rates.

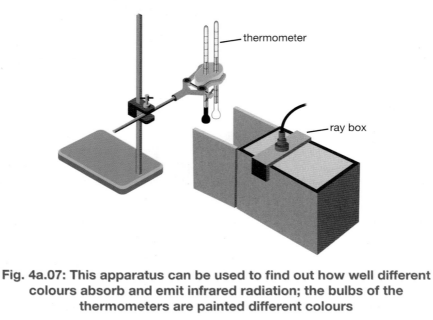

Fig. 4a.07: This apparatus can be used to find out how well different colours absorb and emit infrared radiation; the bulbs of the thermometers are painted different colours

Convection in everyday life

Convection happens around us, and is useful in everyday life. Some examples of convection are:

- in cooking, it helps to transfer energy to all of the contents of a saucepan
- it spreads heat energy from a radiator all around a room
- it creates thermals used by glider pilots, but can also help to create thunderstorms
- it creates sea breezes during the day and land breezes at night (Fig. 4a.08).

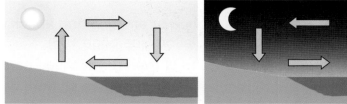

sea breezes – air above the land is warmer than air above the sea

land breezes – air above the sea is warmer than air above the land

Fig. 4a.08: Land and sea breezes occur because the land heats up faster than the sea during the day, and cools down faster at night

Reducing energy transfers

Insulating materials are used to reduce heat transfers.

Most insulating materials are made from non-metals and contain pockets of trapped air. Air is a very poor conductor of heat. Air transfers heat by convection quite well, but convection cannot happen if the air cannot move around.

Some insulating materials use shiny metal foil, which reflects radiated heat, or include a vacuum which cannot transfer heat by conduction or convection.

Humans need to maintain a constant body temperature, and one way in which we do this is to wear clothing. Some ways in which the human body is kept warm are:

- wearing fleece tops or jumpers made of wool or similar materials which contain lots of trapped air
- using duvets on beds, which are filled with tiny feathers or artificial materials that trap air
- using shiny foil blankets to wrap marathon runners or other athletes when they stop exercising, so they do not cool down too quickly.

Insulation is also used in buildings to avoid wasting too much heat energy in cold weather. Buildings can be insulated by:

- stopping draughts around doors and windows
- fitting double-glazed windows
- filling the gap between two layers of a wall with foam or other material (cavity wall insulation)
- covering the floor in the loft with fluffy material
- sticking metal foil to the wall behind radiators.

You should now be able to:

* ★ name some different forms of energy (see page 100)
* ★ state that energy is conserved (see page 101)
* ★ calculate the efficiency of energy transfers (see page 101)
* ★ draw Sankey diagrams to represent energy transfers (see page 103)
* ★ explain how energy is transferred by conduction, convection and radiation (see page 104)
* ★ describe the kinds of surface that are (a) good at emitting and absorbing radiation, and (b) poor at emitting and absorbing radiation (see page 106)
* ★ describe some uses or effects of convection in everyday life (see page 107)
* ★ describe how insulation is used to reduce energy transfers (a) from buildings, and (b) from people (see page 107).

Review questions

1. A motor transfers 30 J of useful energy and wastes 20 J of energy as heat and sound. What is the efficiency of the motor?

2. Draw a Sankey diagram to represent the energy transfer in question 1.

3. Which process of heat transfer is mainly responsible for:

   (a) cooking food under a grill

   (b) warm air from a radiator spreading around a room

   (c) a saucepan handle becoming hot

   (d) feeling warm when you stand in the Sun?

Practice questions

1. A model steam engine runs off small blocks of chemical fuel. It is used to lift a weight.

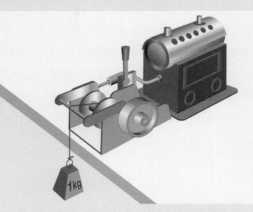

(a) Draw a flow diagram to show the energy transfers involved in this process. **(4)**

The steam engine is replaced with an electric motor. It uses 7 J of electrical energy to give 5 J of gravitational potential energy to the weight.

(b) What is the efficiency of the motor? **(2)**

(c) Draw a Sankey diagram to represent the energy changes when the motor lifts the weight. **(2)**

2. The diagram shows a design for a simple solar cooker.

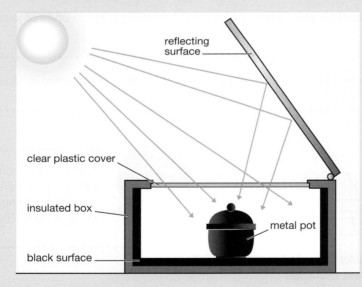

(a) Suggest why the inside of the box and the cooking pot are painted black. **(1)**

(b) (i) How does thermal energy pass from the hot interior of the box into the food inside the pot? Explain your answer. **(1)**

    (ii) Why is the pot made from metal? **(1)**

(c) Which method of heat transfer is the transparent plastic lid designed to reduce? Explain your answer. **(3)**

(d) (i) Which method of heat transfer is the insulated box designed to reduce? **(1)**

    (ii) Suggest what type of material could be used to insulate the box, and explain your suggestion. **(2)**

B Work and power

You will be expected to:

★ recall and use the formula for work, force and distance
★ understand that work done is equal to the amount of energy transferred
★ recall and use the formulae for gravitational potential energy and kinetic energy
★ link gravitational potential energy, kinetic energy and work
★ define power and use the formula for work, power and time.

Work

Moving objects carry on moving unless they are acted on by a force. On Earth there is always friction between a moving object and its surroundings, so a force is needed to keep something moving.

Work is the amount of energy needed to move something for a given distance. The larger the force and the distance, the greater the amount of work done.

Work is calculated using the force and the distance moved in the direction of the force:

$$\text{work done} = \text{force} \times \text{distance moved}$$
$$W = F \times d$$

Because work is a way of describing the amount of energy transferred, it has units of joules (J).

CAM

This can also be written as $\Delta W = Fd = \Delta E$, where the Δ (the Greek letter delta) means *change in*.

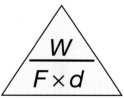

Fig. 4b.01: Formula triangle for work, force and distance

TIP You need to remember this formula as it will not be given to you in the exam. You also need to be able to rearrange it.

Work is calculated from the distance moved *in the direction of the force*. If you carry a heavy bag 100 m across a flat floor you are doing no work. You are exerting a force to stop the bag falling to the ground, but that force is at right angles to the direction in which you have moved the bag. You have done no work, when using 'work' in its physics sense.

Worked example

A boy pushes a box weighing 40 N across the floor. He moves it 5 m, and uses a force of 20 N. What work has he done?

Answer

For this question, the weight of the box does not matter, so ignore the 40 N.

$$\text{work} = 20\text{ N} \times 5\text{ m}$$
$$= 100\text{ J}$$

Gravitational potential energy

It takes energy to move something upwards, as gravity is trying to pull it downwards. So, something moved into a higher position stores the energy used to put it there. This stored energy is called **gravitational potential energy**, and is calculated using the following formula:

$$\text{gravitational potential energy} = \text{mass} \times g \times \text{height}$$
$$GPE = m \times g \times h$$

The weight of an object is its mass $\times$ g. So this formula is just another version of the formula for work, where the force is the weight of the object, and the distance is the height through which it has been lifted.

Fig. 4b.02: Formula triangle for gravitational potential energy

TIP You need to remember this formula as it will not be given to you in the exam. You also need to be able to rearrange it.

Worked example

A woman lifts a 1 kg tin of paint from the floor onto a shelf 1.5 m above the floor. How much gravitational potential energy does the tin of paint now have? Use a value of 9.81 N/kg for g.

Answer

$$GPE = m \times g \times h$$
$$= 1\text{ kg} \times 9.81\text{ N/kg} \times 1.5\text{ m}$$
$$= 14.7\text{ J}$$

Kinetic energy

Moving objects have kinetic energy. The amount of kinetic energy depends on the mass of the object and on the square of its velocity:

$$\text{kinetic energy} = \frac{1}{2} \times \text{mass} \times \text{speed}^2$$

$$KE = \frac{1}{2} \times m \times v^2$$

For example, if the speed of an object doubles, its kinetic energy goes up by a factor of 4.

TIP You need to remember this formula as it will not be given to you in the exam. You also need to be able to rearrange it.

Worked example

A man is cycling at 9 m/s. The total mass of the man and his bike is 90 kg. What is their kinetic energy?

Answer

$$\text{kinetic energy} = \frac{1}{2} \times \text{mass} \times \text{velocity}^2$$

$$= \frac{1}{2} \times 90 \text{ kg} \times 9 \text{ m/s} \times 9 \text{ m/s}$$

$$= 3645 \text{ J}$$

You may be asked to rearrange this formula in an exam. It is not as easy to rearrange as most of the other formulae you will use in this course. If you are not confident with algebra, you could try to memorise the following versions of the formula.

$$m = \frac{2 \times KE}{v^2} \qquad v = \sqrt{\frac{2 \times KE}{m}}$$

TIP If you are asked to work out a velocity using the kinetic energy formula, do not forget to find the square root at the end of your calculation.

Changes in height and speed

Objects that move up and down are usually transferring energy between two different forms. As energy is conserved, the total amount of energy remains the same. A pendulum is a simple example of this.

In Fig. 4b.03 the pendulum is at its highest points at A and E, where it stops before swinging the other way. It is moving fastest at C, which is also its lowest point.

- GPE is maximum at A, KE is zero.
- At B, GPE is being converted to KE.
- GPE is zero at C and KE is maximum.
- KE is being converted to GPE at D
- GPE is maximum again at E, KE is zero.

The pendulum does not keep swinging forever, because air resistance slows it down during each swing. So, not all of the GPE is converted to KE and back, as some of it is converted to heat in the surrounding air while the bob is moving. However, the total amount of energy remains the same.

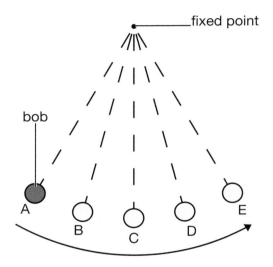

**Fig. 4b.03: A pendulum continually exchanges
gravitational potential energy and kinetic energy**

Power

Power is the rate of transfer of energy, or the rate of doing work. It is measured in watts (W), where 1 watt is 1 joule of energy being transferred each second. Two machines may transfer the same amount of energy, but a more powerful machine will transfer the energy in a shorter time.

Power, work and time are related by the following formula:

$$\text{power} = \frac{\text{work done}}{\text{time taken}}$$

$$P = \frac{W}{t}$$

CAM

This can also be written as:

$$\text{power} = \frac{\text{energy transferred}}{\text{time taken}}$$

$$P = \frac{E}{t}$$

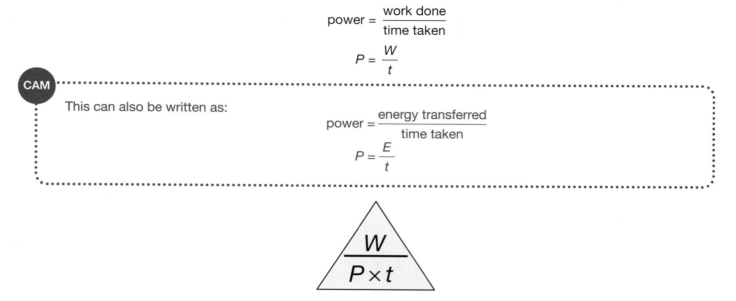

Fig. 4b.04: Formula triangle for work, power and time

Worked example

A woman runs up a flight of steps in 4 seconds. There are 13 steps, and each one has a height of 20 cm. Her mass is 60 kg. What is her power?

Answer

First, find the total height the woman has gained, in metres.

$$\text{height gain} = 13 \times 0.2 \text{ m}$$
$$= 2.6 \text{ m}$$

The energy transferred is the gravitational potential energy the woman has gained by running up the stairs.

$$\text{GPE} = m \times g \times h$$
$$= 60 \text{ kg} \times 9.81 \text{ N/kg} \times 2.6 \text{ m}$$
$$= 1530.4 \text{ J}$$

$$\text{power} = \frac{\text{work done (or energy transferred)}}{\text{time taken}}$$
$$= \frac{1530.4 \text{ J}}{4 \text{ s}}$$
$$= 382.6 \text{ W}$$

You should now be able to:

★ recall and use the formula for work, force and distance (see page 110)
★ understand that work done is equal to the amount of energy transferred (see page 110)
★ recall and use the formulae for gravitational potential energy and kinetic energy (see pages 111 and 112)
★ link gravitational potential energy, kinetic energy and work (see page 113)
★ define the term power and use the formula for work, power and time (see page 114).

Review questions

Use a value of 9.81 N/kg for *g* in these questions.

1. A force does 500 J of work when it is used to move an object 2.5 m. How big is the force?

2. A 5 kg mass is lifted through a distance of 10 m. What is the gravitational potential energy of the lifted mass?

3. A 10 kg mass is given 300 J of gravitational potential energy. How far is it lifted?

4. A 2 kg object is moving at 5 m/s. What is its kinetic energy?

5. A moving object has 100 J of kinetic energy and a mass of 0.5 kg. How fast is it moving?

6. 300 J of work is done in 2 seconds. What power does this represent?

Practice questions

Use a value of 10 N/kg for *g* in these questions.

1. A woman pushes a pushchair and baby 20 m along a ramp to a footbridge. A man carries a suitcase up a set of steps to the same bridge, covering a horizontal distance of 4 m. The total mass of the woman, pushchair and baby is 90 kg – the same as the total mass of the man and suitcase. The top of the bridge is 6 m above the ground.

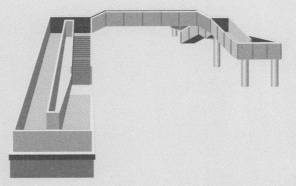

(a) Who has done the most work? Explain your answer. **(3)**

(b) How much work does the man do as he carries the suitcase onto the bridge? **(5)**

(c) How much gravitational potential energy has he gained when he reaches the top? Explain why you can give this answer without carrying out a calculation. **(2)**

(d) He takes 30 seconds to climb the steps. What is his power? **(3)**

2. The drawing shows a track for toy racing cars. The top of the track is 1.5 m above the floor. Each toy car has a mass of 50 g (0.05 kg).

(a) What is the GPE of a toy car at the top of the track? (3)

(b) What is the kinetic energy of the car when it reaches the bottom of the slope? Ignore any energy transfers due to friction. (1)

(c) What is the speed of the car at the bottom of the slope? (3)

(d) The cars go around the loops in the track. Explain why the speed of a car at the top of the loop will be less than at the bottom. (3)

C Energy resources and electricity generation

You will be expected to:

★ understand the energy transfers involved in generating electricity using different energy resources

★ understand the difference between renewable and non-renewable energy resources

★ describe the advantages and disadvantages of generating electricity from different resources.

Generating electricity

Most of the electricity we use is generated by spinning a coil of wire in a magnetic field (see Section 6C). In power stations, the energy for the spinning is transferred from energy stored in fossil fuels or in uranium. The energy can also come from renewable energy resources, such as the wind or waves.

TIP Remember that electricity is not a *source* of energy. It is a very convenient way of transferring energy from one place to another, but it must be produced using other forms of energy.

Electricity can also be produced directly, using solar cells. When light energy falls on the materials in solar cells they produce electricity. Solar cells produce direct current. Most satellites use solar cells to produce electricity.

Renewable and non-renewable resources

Fossil fuels (coal, oil and gas) are **non-renewable** energy resources. Although fossil fuels are probably being formed beneath the surface of the Earth somewhere in the world, we are using them at a *much* faster rate than they are being formed. Fossil fuels will run out one day.

Nuclear power stations use energy stored in **uranium-235** (see Section 7B). Although supplies of **nuclear fuel** will last much longer than supplies of fossil fuels, they will also run out one day. Nuclear fuels are non-renewable.

Renewable resources are energy resources that will not run out. They include:

*   energy transferred by wind
*   energy transferred by moving water
*   solar energy
*   geothermal energy.

Some renewable resources can be used directly, as well as being used to generate electricity. For example, solar energy and geothermal energy can be used to heat water, which can be used to heat buildings.

What makes a good energy resource?

A good energy resource:

*   is cheap
*   has no harmful effects on the environment
*   is available all the time
*   is capable of producing enough electricity for our needs
*   allows fast changes in the amount of electricity produced.

None of the energy resources we use meet all these criteria.

Electricity from fossil fuels and nuclear power

Fossil fuel and nuclear power stations convert energy stored in their fuel into heat energy. This is used to heat water to make high-pressure steam. The steam is directed onto large turbines and makes them spin. The spinning turbines drive generators which produce electricity. A store of chemical energy (or nuclear energy) is being transferred to heat energy, then to kinetic energy and finally to electrical energy.

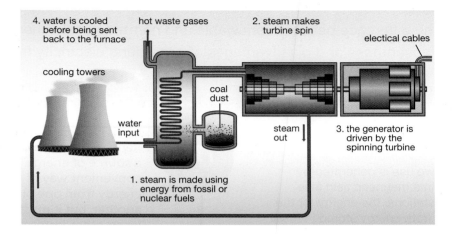

Fig. 4c.01: How a power station works: see Section 7B (page 186) for a diagram of a nuclear reactor

Fossil fuels and nuclear fuel are non-renewable resources. However, electricity produced using fossil fuels is cheaper than electricity produced using other methods. Electricity produced using non-renewable fuels is available all the time. However, it takes some time to start and stop these power stations. So they are left running even when the demand for electricity is low, and some energy is usually wasted.

Burning fossil fuels produces carbon dioxide, which is contributing to global climate change. Sulfur dioxide in the waste gases contributes to acid rain. Sulfur dioxide is now removed from the waste gases from power stations in Europe and other parts of the world, but acid rain is still a problem in some industrialising countries.

Nuclear power stations do not add carbon dioxide to the atmosphere, but the safe disposal of radioactive waste is difficult (see Section 7A), and many people are worried about the possible consequences of an accident at a nuclear power station.

Advantages and disadvantages

Fossil fuels

✓ relatively cheap
✗ polluting
✓ available all the time
– can produce enough electricity for our needs at present, but supplies will eventually run out

Nuclear energy

✗ expensive to build and decommission power stations
– does not produce carbon dioxide, but disposal of waste is difficult
✓ available all the time
– could produce enough electricity for our needs if enough new power stations were built, and nuclear fuel will last much longer than fossil fuels

Fossil fuels are the most widely used energy resource, as they are cheap and the technology to use them is well developed. Nuclear power provides a lot of the electricity in some countries, such as France, but many countries limit the number of nuclear power stations because people are worried about safety, and because it is very expensive to build the power stations.

Energy from the Sun (solar energy)

Solar cells are used to generate electricity directly. Solar energy can also be used to run larger-scale power stations.

One way of doing this is to use a large array of steerable mirrors to focus sunlight onto a central furnace. This gets very hot, and the high temperatures are used to turn water into steam, which turns turbines and generators.

Solar energy can also be used in a 'solar tower' power station. This is like a big chimney with a large area of glass around its base. This type of power station can continue to produce electricity for some time after the Sun stops shining, as it takes time for the ground beneath it to cool down.

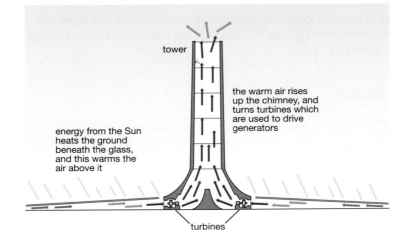

Fig. 4c.02: A solar tower power station

In both cases, heat energy from the Sun is being converted to kinetic energy in the turbines and then to electrical energy.

Energy from the wind

Wind turbines are used to generate electricity from the kinetic energy in the wind. A small generator is mounted at the top of the tower. Wind turbines cannot produce electricity if the wind is not blowing hard enough. They have to be shut down if the wind is too strong, for safety reasons.

Some people object to the building of wind turbines because they spoil the view, particularly in remote, scenic areas.

Wind turbines can be placed out at sea, but they cost more to build and are more difficult to maintain.

Electricity from waves

Waves on the sea can be used to generate electricity, although ways to do this are still being developed. One method is to build a tube on the shore with a pipe leading up to a turbine. Waves push air up and down the pipe, and the moving air makes a turbine spin.

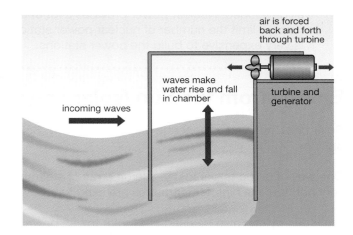

Fig. 4c.03: Energy in waves can be used to generate electricity by pushing air past a turbine

Another way of using waves is to have long lines of floats anchored on the sea. As the waves pass, they make the floats bob up and down. This movement drives tiny generators inside the floats. In both cases, kinetic energy is being transferred to electrical energy.

Electricity from tides

The **tides** can also be used to generate electricity by building a **barrage** (a large dam) across a river estuary. Gates in the barrage are closed at high tide. When the tide has fallen, the gates are opened and trapped water flows out through the turbines.

There are not very many river estuaries in the world suitable for building tidal barrages, and many people are concerned about the effects the barrage and the change in water flow would have on wildlife.

Kinetic energy in tidal currents can also be transferred to electricity using **tidal stream turbines**. These are a bit like underwater wind turbines. They have to be placed on parts of the sea bed where tidal currents are very fast, and so they are harder to build than wind turbines. Tidal stream turbine technology is still being developed.

Advantages and disadvantages

Solar, wind, waves, tidal

✗   expensive

✓   not polluting, although there are some concerns about the effects on wildlife

✗   wind and waves depend on the weather, solar depends on the time of day and the weather, tidal is only available at certain (but predictable) times

✗   suitable locations for all these resources are limited

TIP Energy from the tides does not depend on the weather, but it is only available at certain times of day. These times change, but the times can be predicted many years ahead.

CAM

Solar energy is used in many hot countries for heating water. It is not yet widely used for generating electricity because the technology to produce electricity efficiently is still being developed. Wave and tidal energy can only be used by countries with suitable coastlines. Wind energy is used in many places, but only provides a very small proportion of the electricity needed.

Hydroelectricity

Hydroelectricity is generated using moving water in rivers, or running downhill from reservoirs. Hydroelectric power stations can be switched on and off very quickly, so they are useful for meeting sudden demands for electricity.

Hydroelectric power stations can also be used to store energy. **Pumped storage** power stations have two reservoirs. When water runs down from the top to the bottom reservoir, they are operating in the same way as normal power stations. At night, when the demand for electricity is low but fossil-fuel power stations are still generating, the 'spare' electricity is used to pump water back up to the top reservoir. The energy is stored as gravitational potential energy.

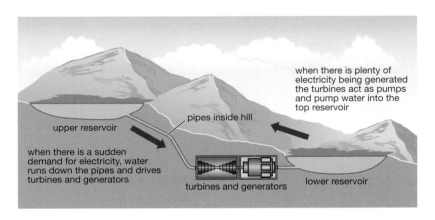

Fig. 4c.04: A pumped storage power station

Advantages and disadvantages

Hydroelectricity
- ✓ relatively cheap
- ✓ some release of polluting gases from new reservoirs – minor compared with fossil fuels
- ✓ available all the time
- ✗ suitable rivers or places to build reservoirs are limited

Geothermal resources

Geothermal energy is widely used in places like Iceland, which have very hot rocks close to the surface. Holes are drilled into the ground, and water is pumped down. This is heated by the rocks and produces steam, which is used to turn turbines. The hot water can also be used for heating homes and other buildings.

Not many countries have large areas with rocks hot enough to make a power station work.

Advantages and disadvantages

Geothermal energy
- ✓ relatively cheap
- ✓ some release of polluting gases from rising hot water – minor compared with fossil fuels
- ✓ available all the time
- ✗ suitable locations are limited

CAM

Most countries use hydroelectricity and geothermal energy when they can, but their use is limited because there are not many suitable places to build the power stations.

You should now be able to:

★ explain why fossil fuels are referred to as non-renewable resources (see page 117)

★ explain why energy resources such as solar, wind and hydroelectricity are referred to as renewable resources (see page 117)

★ describe the energy transfers involved in generating electricity using fossil fuels and nuclear fuel (see page 118)

★ describe the advantages and disadvantages of using fossil fuels and nuclear fuel to generate electricity (see page 118)

★ describe how electricity can be generated using (a) solar energy, (b) tides, (c) waves, (d) hydroelectricity (see page 119)

★ describe the advantages and disadvantages of generating electricity from different renewable resources (see page 119).

Review questions

1. Summarise the energy changes that take place in a fossil-fuelled power station.

2. Outline two different ways in which solar energy can be used to spin generators in power stations.

3. Write down the renewable resources that:

   (a) can only be used when the weather is suitable

   (b) can be used at any time

   (c) do not depend on the weather, but are not available all the time.

Practice questions

1. Residents on an island use electricity generated by three wind turbines, and from solar cells on south-facing roof tops. They also have a generator that runs off diesel fuel.

   (a) Why are wind and solar power referred to as renewable resources? **(1)**

   (b) What are the energy transfers involved in generating electricity:
      (i) from the wind **(1)**
      (ii) using the diesel generator? **(1)**

   (c) Why do the islanders need a diesel-powered generator as well as the wind turbines and solar panels? **(3)**

   (d) What are the disadvantages of using diesel fuel to generate electricity? **(3)**

   (e) There is a suggestion that a tidal stream turbine could be installed.
      (i) How will this reduce the need for the diesel generator? **(1)**
      (ii) Why won't this mean that the generator is totally unnecessary? **(1)**

Section Five

5 Solids, liquids and gases

A Density and pressure

You will be expected to:

★ recall and use the formula relating density, mass and volume
★ describe how to find density using measurements of mass and volume
★ recall and use the formula relating pressure, force and area
★ understand that pressure in a liquid or gas acts in all directions
★ recall and use the formula relating pressure, height and density of a fluid.

Units

You will use the following quantities and units in this section.

Quantity	Symbol	Unit
area	A	square metres (m^2)
density	ρ	kilograms/cubic metre (kg/m^3)
force	F	newtons (N)
length or height	l or h	metres (m)
mass	m	kilograms (kg)
pressure	p	pascals (Pa) or N/m^2
volume	V	cubic metres (m^3)

Density

The **density** of a substance is the mass of a certain volume of the substance:

$$density = \frac{mass}{volume}$$

$$\rho = \frac{m}{V}$$

The units for density depend on the units used to measure the mass and volume. The standard unit in science is kg/m^3, but you may also see densities given in g/cm^3. When you use a value for density in other formulae, such as the one for pressure difference in liquids (see page 128), its units must be kg/m^3.

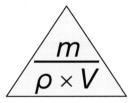

TIP You need to remember this formula as it will not be given to you in the exam. You also need to be able to rearrange it.

Fig. 5a.01: Formula triangle for density, mass and volume

Worked example

A cube of copper measures 3 cm × 3 cm × 3 cm, and has a mass of 0.241 kg. What is its density?

Answer

First find the volume in m^3.

$$3\text{ cm} = 0.03\text{ m}$$

$$V = 0.03\text{ m} \times 0.03\text{ m} \times 0.03\text{ m}$$

$$= 2.7 \times 10^{-5}\text{ m}^3$$

$$density = \frac{mass}{volume}$$

$$= \frac{0.241\text{ kg}}{2.7 \times 10^{-5}\text{ m}^3}$$

$$= 8926\text{ kg/m}^3$$

Measuring volumes to find densities

You need to know the mass and the volume of an object to find its density.

If the object is a cube or other regular shape, the volume can be found by multiplying its dimensions. The standard unit for volume in science is m^3, but for small objects it is often more convenient to use cm^3 or mm^3. Whichever unit you use, all three dimensions of an object must be in the same unit.

The volume of irregular shapes can be found by **displacement**.

- If the object is small enough to fit inside a measuring cylinder, half-fill the measuring cylinder with water and write down the volume of water.
- Carefully drop the object into the water, or suspend it on a cotton thread, and read the volume again.
- The difference between the two volumes is the volume of the object.

If the object is too big to fit in a measuring cylinder you need to use a **displacement can** (Fig. 5a.02). This is a can with a spout in the side.

- The can is filled with water to the level of the bottom of the spout.
- The object is lowered into it on a thin thread.
- The volume of water that runs out of the spout is measured.
- The volume of water is the same as the volume of the object.

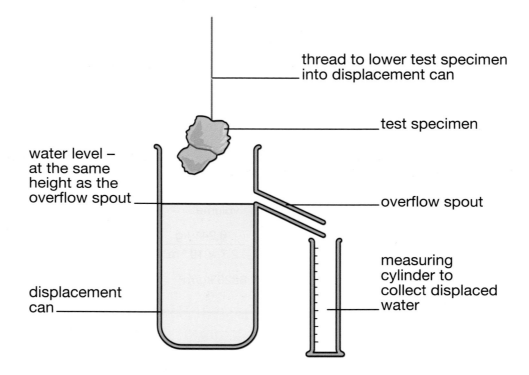

thread to lower test specimen into displacement can

test specimen

water level – at the same height as the overflow spout

overflow spout

measuring cylinder to collect displaced water

displacement can

Fig. 5a.02: Using a displacement can to measure the volume of an irregular solid

Pressure

Pressure is the force on a certain area:

$$\text{pressure} = \frac{\text{force}}{\text{area}}$$

$$p = \frac{F}{A}$$

For example, you have the same weight when using ice skates or snowshoes, but the pressure you exert on the ground is much higher in ice skates because the area of contact with the ground is much smaller.

Pressures can be given in N/cm^2, but the correct scientific units to use are pascals (Pa). 1 Pa = 1 N/m^2.

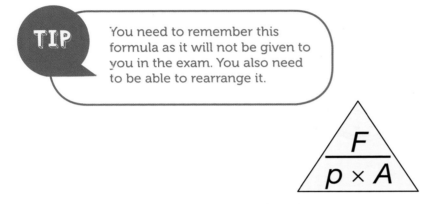

TIP You need to remember this formula as it will not be given to you in the exam. You also need to be able to rearrange it.

$$\frac{F}{p \times A}$$

Fig. 5a.03: Formula triangle for pressure, force and area

Worked example

A walker and his rucksack have a total weight of 800 N. The pressure he exerts on the ground when standing on both feet is 20 kPa. What area do his boots cover?

Answer

$$\text{area} = \frac{\text{force}}{\text{pressure}}$$

$$= \frac{800 \text{ N}}{20\ 000 \text{ Pa}}$$

$$= 0.04 \text{ m}^2$$

TIP Don't forget to convert the pressure into pascals before using the number in the formula.

Pressure in fluids

Liquids and gases are both referred to as **fluids**. The pressure in a fluid acts equally in all directions.

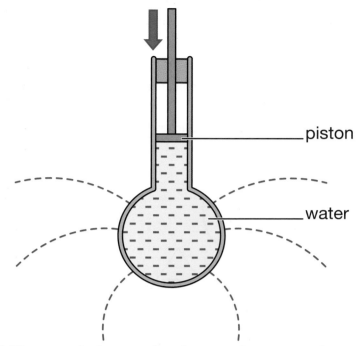

Fig. 5a.04: The water in the container is put under pressure by pushing the piston; the water squirts out through the holes equally in all directions

The deeper the fluid, the greater the weight of fluid above that is contributing to the pressure, so the pressure in a fluid increases with depth. The pressure also depends on the density of the fluid.

The pressure difference between two points in a fluid can be calculated using this formula:

$$\text{pressure difference} = \text{height} \times \text{density} \times g$$
$$p = h \times \rho \times g$$

The value for density must be in kg/m³. The value for g is 9.81 N/kg, but exam papers usually tell you to use a value of 10 N/kg.

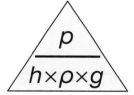

Fig. 5a.05: Formula triangle for pressure difference in a fluid

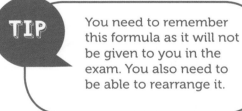

TIP

You need to remember this formula as it will not be given to you in the exam. You also need to be able to rearrange it.

Worked examples

Example 1

A simple mercury barometer (Fig. 5a.06) consists of a sealed tube filled with mercury, inverted into an open dish of mercury. The pressure from the mercury at the base of the column is equal to the pressure of the atmosphere on the mercury in the dish.

The density of mercury is 13 534 kg/m³. If the column of mercury is 0.77 m high, what is atmospheric pressure? Use g = 10 N/kg.

Answer

pressure = height × density × g

= 0.77 m × 13 534 kg/m³ × 10 N/kg

= 104 212 Pa

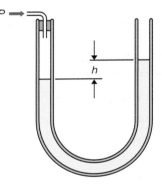

Fig. 5a.06: A mercury barometer

Example 2

A manometer is a device for measuring gas pressure. It consists of a U-shaped tube filled with liquid. One arm of the tube is connected to the gas supply, and the other is left open. The difference in height of the liquid in the two sides of the tube is a measure of the pressure of the gas above atmospheric pressure.

The manometer in Fig. 5a.07 shows a difference in height of 5 cm when measuring a pressure of 400 Pa. What is the density of the liquid in the manometer?

Answer

h = 5 cm = 0.05 m

$$\text{density} = \frac{\text{pressure}}{\text{height} \times g}$$

$$= \frac{400 \text{ Pa}}{0.05 \text{ m} \times 10 \text{ N/kg}}$$

$$= 800 \text{ kg/m}^3$$

Fig. 5a.07: A manometer

You should now be able to:

★ recall and use the units and symbols for density and pressure (see page 124)
★ recall and use the formula relating density, mass and volume (see page 125)
★ describe how to find density using measurements of mass and volume (see page 126)
★ recall and use the formula relating pressure, force and area (see page 127)
★ describe how the pressure in a liquid or gas acts in all directions (see page 128)
★ recall and use the formula relating pressure, height and density of a fluid
    (see page 128).

Review questions

1. An object has a volume of 0.05 m³ and a density of 1200 kg/m³. What is its mass?

2. What is the area of an object with a weight of 60 N that exerts a pressure of 50 Pa on the surface beneath it?

3. What is the density of a fluid that exerts a pressure of 8000 Pa beneath a column 0.5 m high? (Use g = 10 N/kg.)

Practice questions

1. A steel drum has a mass of 10 kg when it is empty. Its radius is 0.25 m and its height is 1 m. It is used to store heating oil. When it is full of oil it has a mass of 180 kg. (Use g = 10 N/kg.)

    (a) The volume of a cylinder is given by $V = \pi r^2 h$. What is the volume of oil in the barrel?   **(1)**

    (b) What is the density of the oil?   **(3)**

    (c) What is the pressure on the bottom of the barrel caused by the oil above it?   **(2)**

    (d) What pressure does the bottom of the barrel exert on the ground beneath it?   **(5)**

B Change of state

CAM ★ describe the properties of solids, liquids and gases

★ describe how the particles move and are arranged in solids, liquids and gases

★ describe what happens in melting, evaporation and boiling

CAM ★ explain why substances expand when they are heated, and some practical effects of this

★ describe different ways of measuring temperature

★ explain what thermal capacity means

★ describe an experiment to measure the specific heat capacity of a substance

★ explain what latent heat is in terms of particles

★ describe an experiment to measure latent heats

★ describe the factors that affect the rate of evaporation, and why evaporation leads to cooling.

CAM

Solids, liquids and gases

Solids, liquids and gases have different properties.

Solids:

- keep a fixed volume
- keep a fixed shape, so cannot be poured
- are difficult to compress.

Liquids:

- keep a fixed volume
- take the shape of their container, and can be poured
- are difficult to compress.

Gases:

- do not have a fixed volume, and expand to fill their container
- do not have a fixed shape, and take the shape of their container
- are easy to compress.

Particles in solids, liquids and gases

The properties of solids, liquids and gases can be explained by the way the particles in them are arranged. The particles in a substance can be atoms or molecules, depending on the substance.

solid   liquid   gas

Fig. 5b.01: Particles in solids, liquids and gases

In a solid the particles:

*   are close together
*   are held by strong forces in a regular, fixed arrangement
*   can only vibrate about their fixed positions.

In a liquid the particles:

*   are close together, but are in an irregular arrangement
*   are held together by fairly strong forces
*   move around randomly, but stay within the liquid.

In a gas the particles:

*   are far apart
*   have only very weak forces between them
*   move around freely, in random directions.

The closeness of the particles in solids and liquids explains why substances in these states are hard to compress.

The strength of the bonds between particles in solids explains why they have fixed shapes. The lack of strong bonds in gases explains why gases expand to fill their containers.

Changing state

Substances can change from one state to another when the thermal energy in them changes.

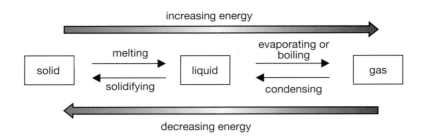

Fig. 5b.02: Changes of state

Evaporation can happen at any temperature. The hotter a liquid is, the faster the evaporation rate. When a liquid reaches its boiling point, evaporation happens as quickly as possible.

TIP Be careful to use *evaporation* and *boiling* correctly if you are answering a question about changes of state.

While a substance is changing state its temperature does not change, even though the substance is still being heated or cooled. Fig. 5b.03 shows how the temperature changes when substances are heated or cooled.

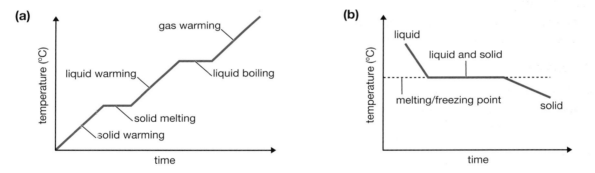

**Fig. 5b.03: (a) Heating curve for a solid being heated to above its boiling point
(b) Cooling curve for a substance that is solidifying**

TIP Remember that energy is still being transferred when a material is changing state.

Thermal expansion

Most substances **expand** when they are heated, because the extra internal energy makes the particles move faster.

- For solids, the extra movement is an increased vibration about a fixed point.
- In liquids and gases, the particles move around faster.

When a substance is cooled its particles have less energy and take up less space, so the substance **contracts**.

In general, a particular material will expand more per degree of temperature change when it is a liquid than when it is a solid. Gases expand even more for a given temperature change.

Thermal expansion needs to be taken into account when designing many objects.

- Bridges and other structures are designed and built to make allowances for the material to expand in warm weather. Without this, parts of the structure would bend when the temperature increases and the material expands.
- Engines are not very efficient when they are first started up because there are gaps between some of the components. When the engine reaches its normal operating temperature, the components have expanded and all the parts fit together better.
- Electricity and telephone wires are not stretched tight between their poles. If they were, they would break in the winter when they got cold and contracted.

Measuring temperature

A temperature scale is defined between two fixed points. These are temperatures that do not change. For the Celsius scale, the two fixed points are the melting and boiling points of pure water (0 °C and 100 °C).

Measuring temperature uses different properties of materials. Any physical property that is used in measuring temperature needs to show **linearity**. This means that the change in the property per degree of temperature change is the same whatever the temperature.

The **range** of a thermometer is the spread of temperatures it can read. Thermometers used in school science labs often have a range of –10 °C to 110 °C.

The **sensitivity** of a measuring instrument describes the smallest changes that it can detect. A long thermometer with the same range as a short one is likely to be more sensitive, because there will be a bigger gap between each degree mark.

Thermometers

Liquid-in-glass thermometers contain a small amount of liquid. The liquid expands up a narrow tube when its temperature rises, and the temperature is read from a scale.

This type of thermometer makes use of the fact that the volume of the liquid inside it varies with temperature. A liquid-in-glass thermometer shows linearity because the liquid expands the same amount between 10 and 11 °C as it does between 90 and 91 °C.

the bulb of the thermometer contains a reservoir of the liquid

scale

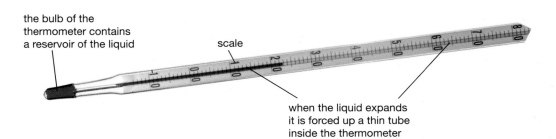

when the liquid expands it is forced up a thin tube inside the thermometer

Fig. 5b.04: The liquid in this liquid-in-glass thermometer is alcohol, with a little dye added to make it easier to see

Liquid-in-glass thermometers are not suitable for measuring temperatures that change quickly because it takes time for the thermometer to reach the same temperature as the material whose temperature it is measuring.

Thermistors

The resistance of a **thermistor** varies with temperature (see page 57), and so a thermistor circuit can be used to measure temperature.

Thermocouples

A **thermocouple** consists of wires of two different metals connected as shown in Fig. 5b.05. If the two junctions are at different temperatures there will be a small voltage. This voltage can be detected with a millivoltmeter.

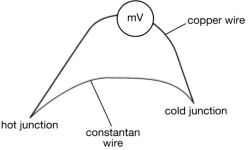

mV

copper wire

cold junction

hot junction

constantan wire

Fig. 5b.05: A thermocouple

To use a thermocouple as a thermometer, one junction (called the **cold junction**) has to be kept at a constant temperature, and the other junction (the **hot junction**) is put into contact with the material whose temperature is being measured.

Thermocouples are useful because they can measure very high temperatures. They can also measure temperatures that are changing very rapidly.

Thermal capacity

The **thermal capacity** (sometimes called the **heat capacity**) of an object is the amount of energy it takes to raise the temperature of the object by 1 °C.

The thermal capacity of an object depends on the material it is made from and on how much of that material there is. So, a large mug of water has a higher thermal capacity than a small cup of water, just because the mug contains more water.

A more useful measure is the **specific heat capacity** of a material. In physics, *specific* usually means a quantity per kilogram. In this case, the specific heat capacity of a material is the amount of energy needed to raise the temperature of 1 kg of the material by 1 °C.

Specific heat capacity can be measured using the method shown in Fig. 5b.06. The energy supplied to the immersion heater is calculated from the current and the voltage, and this is divided by the mass of the block and by the temperature rise.

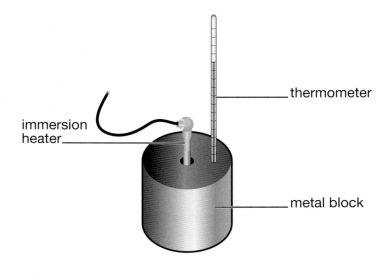

thermometer

immersion
heater

metal block

Fig. 5b.06: Measuring the specific heat capacity of a metal block

Latent heats

The temperature of a solid being heated stops rising when it reaches its melting point (see Fig. 5b.03).

If heating continues, the energy is used to break the bonds between the particles in a solid and allow it to turn into a liquid. Only when all the solid has melted does the temperature start to rise again.

The energy that is used to melt the solid is called the **latent heat of fusion**.

If a liquid is cooled, the temperature falls until it reaches its freezing point. The temperature remains constant while the liquid is turning into a solid, because the energy that was used to break the bonds in the solid is being released as the bonds re-form. The latent heat of fusion is given out when the liquid solidifies.

A similar thing happens when a liquid turns into a gas. The energy needed to break the bonds in the liquid is the **latent heat of vaporisation**. This energy is given out again when the gas condenses back to a liquid. The **specific latent heat** of fusion or vaporisation is the latent heat for 1 kg of the substance.

You can measure the latent heat of vaporisation of steam by heating water on a balance using an electric immersion heater.

- When the water reaches boiling point, find the mass of the water.
- Find the mass again 5 minutes later, when some of the water has evaporated.
- The specific latent heat is the energy used divided by the mass of water that has evaporated.

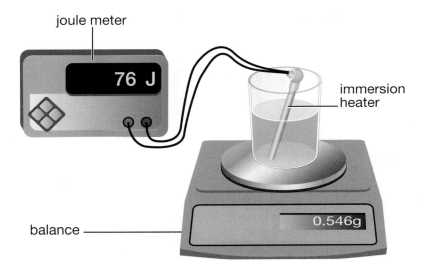

Fig. 5b.07: Apparatus used to measure the specific latent heat of steam

You can use a similar method to find the specific latent heat of fusion for ice. Take ice at 0 °C and measure the energy needed to melt it all. Divide the energy by the mass of ice to find the specific latent heat.

Evaporation and cooling

The particles in a gas move around rapidly, but they do not all move at the same speed. The temperature of a gas is a measure of the *average* speed of the particles.

Evaporation from a liquid can take place at any temperature. When this happens below the boiling point of the liquid, it is the particles with the highest speeds (and so the most energy) that escape. This means that the particles left behind in the liquid have a lower average speed, and so the temperature of the remaining liquid is lower.

Evaporation happens faster:

- at higher temperatures
- with a greater surface area
- with a draught blowing over the surface.

Evaporation happens faster at higher temperatures because more particles in the liquid have enough energy to escape from the liquid and turn into a gas. Also, the greater the surface area of the liquid, the more chance particles have to escape from the liquid.

Air can only hold a certain amount of an evaporated liquid. As the air gets more particles from the evaporating liquid, it becomes harder for more liquid to evaporate. If there is a draught blowing over the liquid, these evaporated particles are moved away, so it is easier for more of the liquid to evaporate.

(a)

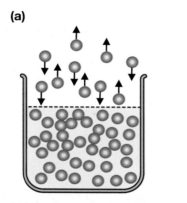

if there is no draught, the evaporated particles collect above the liquid and start to condense again; the overall rate of evaporation is slow

(b)

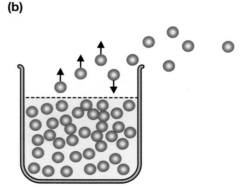

if there is a breeze, the evaporated particles are blown away before they have time to condense again

Fig. 5b.08: (a) In still conditions, the overall rate of evaporation is slow
(b) If there is a breeze, evaporation is faster

You should now be able to:

 ★ describe the properties of solids, liquids and gases (see page 131)

★ describe how the particles move and are arranged in solids, liquids and gases (see page 132)

★ describe what happens in melting, evaporation and boiling (see page 133)

★ sketch a labelled graph showing how the temperature changes as a solid is heated until it becomes a gas (see page 133)

 ★ explain why substances expand when they are heated, and give some practical effects of this (see page 134)

★ describe different ways of measuring temperature (see page 134)

★ explain what thermal capacity means and how it is different from specific heat capacity (see page 136)

★ describe an experiment to measure the specific heat capacity of a substance (see page 136)

★ explain what latent heat is in terms of particles, and how this helps to explain the shape of a graph of temperature against time for a melting or boiling substance (see page 137)

★ describe an experiment to measure latent heats (see page 137)

★ describe the factors that affect the rate of evaporation (see page 138)

★ explain why evaporation leads to cooling (see page 138).

Review questions

1. Which state (or states) of matter does each of these descriptions match?

   (a) particles close together but not fixed

   (b) particles can only vibrate

   (c) no fixed shape

   (d) difficult to compress

   (e) large spaces between particles

2. What is the difference between evaporation and boiling?

3. Sketch a graph showing how the temperature changes as a solid is heated until it becomes a gas. Label the melting and boiling points.

4. Why do metals contract when they are cooled?

5. How do the latent heats of fusion and vaporisation help to explain the shape of the graph you drew for question 3?

6. How does sweat help to cool you down?

Practice questions

1. Two students are having an argument:

   Sand is a solid substance

   Sand is a liquid

   (a) What properties of sand might lead someone to say that it is a liquid? **(2)**

   (b) What properties of sand might lead someone to say that it is a solid? **(2)**

   (c) Describe how the particles move and are arranged in a grain of sand. **(2)**

   (d) Describe the particles in a liquid. **(2)**

   (e) Describe what happens when a solid changes into a liquid. **(2)**

CAM 2. A bimetallic strip is made of two different metals bonded together. The two metals expand by different amounts when they are heated. The diagram shows a bimetallic strip being used as a thermostat to control a heating system.

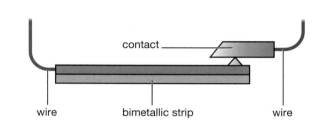

contact

wire   bimetallic strip   wire

   (a) Explain why solid materials expand when they are heated. **(2)**

   (b) The brown metal in the diagram expands more than the grey one for a given temperature change. What will happen when the strip gets hotter? **(1)**

   (c) Explain how the bimetallic strip is used to control a central heating system. **(5)**

   (d) How would the thermostat be changed to control an air conditioning system to keep a room cool? Explain your answer. **(3)**

3. A 'zeer' is a method of keeping food cool in places where there is no electricity to run refrigerators. It consists of two earthenware pots, one inside the other, with the gap between them filled with wet sand. Food is placed inside the inner pot and covered with a damp cloth.

The outer pot must be porous, but the inner one can be glazed to stop water from the sand entering the food. As long as the sand is kept moist and the zeer is placed in a dry place, preferably with a breeze, it can extend the 'shelf life' of food kept in it by up to 3 weeks. The zeer uses evaporation to cool the food.

(a) Explain three factors that affect the rate of evaporation of a liquid. (6)

(b) Explain how the zeer works. (4)

C Ideal gas molecules

Units

You will use the following quantities and units in this section.

Quantity	Symbol	Unit
pressure	p	pascals (Pa) or N/m^2
temperature	T	degrees Celsius (°C), or kelvins (K)
volume	V	cubic metres (m^3)

Brownian motion

Brownian motion is the random 'jiggling' of particles suspended in a fluid. The particles can be bits of dust floating in the air (often seen in a sunbeam or in the beam from a projector), or pollen grains suspended in water.

For dust in air, this motion is explained as the result of much smaller air particles bumping into the dust grains and making them move. Although the air molecules are much, much smaller than the dust grains, they are moving much faster, and so have enough energy to change the direction of movement of the dust.

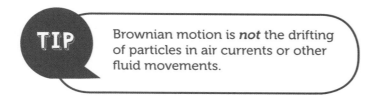

TIP Brownian motion is *not* the drifting of particles in air currents or other fluid movements.

Brownian motion is one piece of evidence that supports the kinetic model of matter (the model that describes how everything is made of particles (see page 132).

The Kelvin temperature scale

The temperature of a gas depends on the average speed of its atoms or molecules. The faster the molecules are moving, the higher the temperature.

If you cool a gas down far enough, the molecules will eventually stop moving. This happens at the same temperature for all substances, at –273 °C. This temperature is called **absolute zero**.

The **Kelvin scale of temperature** starts at absolute zero. The temperature intervals on the Kelvin scale are the same as on the Celsius scale, so on the Kelvin scale ice melts at 273 K and water boils at 373 K.

To convert from the Kelvin scale to the Celsius scale, subtract 273.

To convert from the Celsius scale to the Kelvin scale, add 273.

 **TIP** If you are converting between Celsius and Kelvin temperatures, remember that the Kelvin temperature should always be a higher number than the Celsius temperature, and can never be a negative number.

Worked example

Ethanol freezes at –114 °C. What temperature is this in Kelvin?

Answer

Kelvin temperature = Celsius temperature + 273 K

= –114 + 273

= 159 K

The Kelvin temperature of a gas is proportional to the average kinetic energy of its molecules. So the faster the molecules are moving, the higher the temperature.

Pressure in gases

Molecules in a gas are moving continuously, and bumping into the walls of their container. The force of all these bumps produces a pressure.

If a gas is heated, its molecules move faster and the temperature rises. The faster molecules bump into the walls harder and more often, so the pressure also rises.

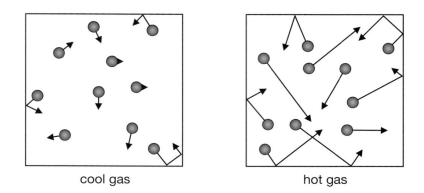

cool gas   hot gas

Fig. 5c.01: Particle motion in gases at different temperatures

If the gas is in a sealed container with a fixed volume, its pressure is proportional to its Kelvin temperature.

Worked example

A container of gas has a pressure of 50 000 Pa at a temperature of 200 K. What is the pressure of the gas at 400 K if the volume does not change?

Answer

The Kelvin temperature has doubled, so the pressure also doubles:

$$\text{pressure} = 100\ 000\ \text{Pa}$$

If a gas is allowed to expand at constant pressure when it is heated, its volume will increase. The volume of a gas at constant pressure is proportional to its Kelvin temperature.

Pressure and temperature calculations

The pressure and temperature of a fixed mass of gas at constant volume are related by this formula:

$$\frac{p_1}{T_1} = \frac{p_2}{T_2}$$

This can be rearranged to give the following relationships:

$$p_2 = \frac{p_1 T_2}{T_1} \qquad T_2 = \frac{p_2 T_1}{p_1}$$

> **Worked example**
>
> A bottle of compressed gas has a volume of 0.2 m³ and contains gas at a pressure of 14 000 000 Pa at 20 °C. What will the pressure be if the temperature rises to 50 °C?
>
> *Answer*
>
> First convert the temperatures to Kelvin: 20 °C = 293 K and 50 °C = 323 K
>
> $$p_2 = \frac{p_1 T_2}{T_1}$$
>
> $$= \frac{14\ 000\ 000\ \text{Pa} \times 323\ \text{K}}{293\ \text{K}}$$
>
> $$= 15\ 000\ 000\ \text{Pa}$$

Pressure and volume calculations

If you compress a gas into a smaller volume, its pressure increases. If you increase the volume, the pressure decreases.

These changes are represented by the following formula:

$$p_1 V_1 = p_2 V_2$$

Note that this formula only applies if the mass of gas does not change and the gas remains at the same temperature.

This can be rearranged to give the following relationships:

$$p_2 = \frac{p_1 V_1}{V_2} \qquad V_2 = \frac{p_1 V_1}{p_2}$$

> **Worked example**
>
> A bicycle pump compresses 0.000 015 m³ of air at atmospheric pressure to a volume of 0.000 005 m³. What is the pressure of the gas when it has been compressed? Take atmospheric pressure to be 101 000 Pa.
>
> *Answer*
>
> $$p_2 = \frac{p_1 V_1}{V_2}$$
>
> $$= 101\ 000\ \text{Pa} \times \frac{0.000\ 015\ \text{m}^2}{0.000\ 005\ \text{m}^2}$$
>
> $$= 303\ 000\ \text{Pa}$$

You should now be able to:

★ describe Brownian motion and explain its significance (see page 142)

★ describe the Kelvin temperature scale and the meaning of absolute zero (see page 143)

★ convert between temperatures on the Kelvin and Celsius scales (see page 143)

★ explain why gases exert pressure on their containers (see page 144)

★ describe how the temperature affects the pressure of a gas when the volume is fixed (see page 144)

★ describe how the volume and pressure of a gas at a fixed temperature are related (see page 145)

★ use the relationship between the pressure and temperature of a fixed mass of gas at a constant volume (see page 145)

★ use the relationship between the pressure and volume of a fixed mass of gas at a constant temperature (see page 145).

Review questions

1. (a) What is 20 °C on the Kelvin scale?

   (b) What is 20 K on the Celsius scale?

2. The temperature of a fixed volume of gas is raised from 280 K to 300 K. Its pressure is 100 000 Pa at 280 K. What is its final pressure?

3. 2 m³ of a gas are compressed to 0.5 m³. If the gas was initially at 50 000 Pa, what is its final pressure?

Practice questions

1. A student measures the pressure of a sample of nitrogen at different temperatures, and plots a graph of her results. She extrapolates the line on her graph until it reaches the horizontal axis.

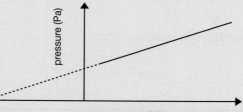

   (a) What causes pressure in a gas? (2)

   (b) Why does the pressure fall if the gas is cooled? (3)

   (c) At what temperature should her line cross the temperature axis? Explain your answer in as much detail as you can. (3)

   (d) The temperature in the lab was 18 °C during the experiment. What is this temperature on the Kelvin scale? (1)

2. A cylinder used to store compressed gases can withstand a pressure of 2.5 × 10⁸ Pa before it explodes. The normal pressure of gas within the cylinder is 5 × 10⁷ Pa when it is stored at 20 °C. At what temperature will it explode? Give your answer in degrees Celsius. (4)

3. The pressure of air in a car tyre is 210 kPa, and the tyre holds 0.001 m³ of air. What volume of air at atmospheric pressure has been put into the tyre? Take atmospheric pressure to be 100 kPa. (3)

Section Six

6 Magnetism and electromagnetism

A Magnetism

You will be expected to:

★ recall that magnets attract magnetic materials, but can attract or repel other magnets
★ explain the difference between magnetically hard and soft materials
★ describe how magnetism can be induced in materials
★ describe some ways of demagnetising an object
★ sketch the magnetic field patterns for a bar magnet and between two bar magnets
★ describe how to use two permanent magnets to produce a uniform magnetic field pattern.

Attracting and repelling

Magnetism is a non-contact force. A magnet can affect other magnets or magnetic materials without having to touch them.

Magnets can affect **magnetic materials**. Magnetic materials include:

- iron and steel
- nickel
- cobalt.

A magnet has two ends, called the **north-seeking pole** and the **south-seeking pole**. These names are usually shortened to **north pole** and **south pole**.

If the north pole of one magnet is placed close to the south pole of another magnet, the two magnets will **attract** each other.

If two north poles, or two south poles, are placed together, the two magnets will **repel** each other.

This behaviour can be summarised as:

- like poles repel
- unlike poles attract.

Properties of magnetic materials

Magnetic materials such as iron and steel can be magnetised, as well as being attracted to magnets. Steel is an alloy of iron with carbon and sometimes some other elements as well. Iron and steel are **ferrous materials**.

Iron is sometimes called a **soft magnetic material**. This does not mean that it is soft to the touch, but that it very quickly loses its magnetism. Steel is a **hard magnetic material**. Once it has been magnetised it keeps its magnetism for a long time.

Cobalt and nickel have similar magnetic properties to iron. They are **non-ferrous materials**.

Magnetising and demagnetising

Some materials become magnetised when they are placed in a magnetic field. We say the magnetism has been 'induced' in the material. For example, an iron nail left in the same position for months in a cupboard will become magnetised by the Earth's magnetic field.

Materials can be magnetised faster by placing them in a stronger magnetic field, such as that produced by a coil of wire with a current flowing through it. Small items like pins can be magnetised by stroking them with a magnet.

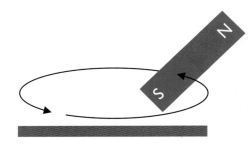

Fig. 6a.01: A magnetic material can be magnetised by stroking it in one direction with a magnet

CAM

Magnetic materials such as iron include many small 'domains'. In unmagnetised iron these domains are arranged in random directions. When the material is placed in a magnetic field, the domains line up and the material itself becomes a magnet.

Magnets can be **demagnetised** (their magnetism can be removed) by heating them or by repeatedly hitting or jarring them. Heating and jarring the magnet both provide enough energy to allow the domains to become randomly oriented again.

A magnet can also be demagnetised using a coil of wire and an alternating current.

- The magnet is placed in the coil, and the current switched on.
- The direction of the magnetic field changes each time the direction of the current changes, and so the alignment of the domains in the magnet also changes.
- If the size of the current is gradually reduced, not all the domains switch direction each time the current changes.
- By the time the current is reduced to zero, all the domains are randomly aligned and the material is no longer a magnet.

Magnetic fields

The space around a magnet where it can affect magnetic materials is its **magnetic field**.

We mark the magnetic field using lines that show which way a freely moving north pole would move (from the north to the south end of the magnet). A magnetic field can be plotted using small compasses called plotting compasses (Fig. 6a.02), or by placing a magnet under a piece of paper and sprinkling iron filings on the paper (see page 152).

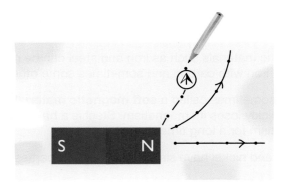

Fig. 6a.02: A plotting compass is used to identify the pattern of field lines by marking the direction in which the needle points at different places around the magnet

Magnetic fields around bar magnets

You need to be able to sketch the magnetic field around a single bar magnet (Fig. 6a.03), and also the field pattern between two bar magnets.

The shape of the field between two bar magnets depends on how they are arranged. Fig. 6a.04 shows the shapes for two magnets placed end to end.

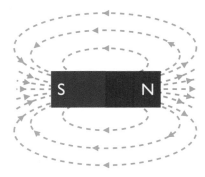

Fig. 6a.03 The magnetic field around a bar magnet

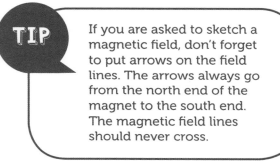

> **TIP** If you are asked to sketch a magnetic field, don't forget to put arrows on the field lines. The arrows always go from the north end of the magnet to the south end. The magnetic field lines should never cross.

The magnetic field between two magnets placed with a north pole near to a south pole provides a **uniform** magnetic field in the area between the two magnets. This means that all the magnetic field lines are parallel to each other, and the strength is the same in the area between the magnets. A uniform magnetic field produced like this is used in simple motors and generators.

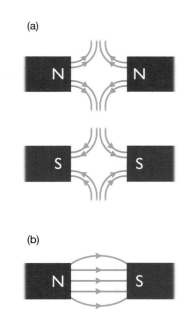

Fig. 6a.04: (a) The magnetic field between two bar magnets placed so that they are repelling each other (b) The magnetic field between two bar magnets placed so that they are attracting each other

You should now be able to:

★ state that magnets attract magnetic materials (see page 148)
★ describe how to arrange magnets so they attract or repel each other (see page 148)
★ explain the difference between magnetically hard and soft materials (see page 149)
★ describe how magnetism can be induced in materials (see page 149)

 ★ describe some ways of demagnetising an object (see page 149)

★ sketch the magnetic field pattern (a) for a single bar magnet, (b) between two bar magnets (see page 150)
★ describe how to use two permanent magnets to produce a uniform magnetic field pattern (see page 150).

Review questions

1. Describe two ways in which magnets can be arranged so that they repel each other.

2. Write down three metals that can be magnetised.

3. What is the difference between hard and soft magnetic materials?

4. How can you magnetise a pin using a bar magnet?

CAM 5. Describe two ways of demagnetising a magnet.

6. Sketch the shape of the magnetic field:

   (a) around a single bar magnet

   (b) between two magnets with their south poles close to each other

   (c) between two magnets with a north pole and a south pole close to each other.

Practice questions

1. A student places a piece of paper over two magnets and sprinkles iron filings onto the paper. The photo shows his results.

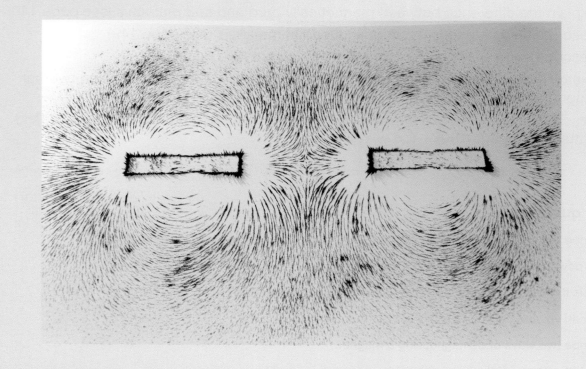

(a) Copy and complete the diagram below to show the shape of the magnetic field. (3)

(b) If the two magnets were free to move, what would happen? (1)

(c) What will happen if one of the magnets is replaced with an unmagnetised piece of iron, and both are free to move? (1)

B Electromagnetism

Electromagnets

A wire with an electric current flowing through it produces a magnetic field around it. A stronger magnetic field can be obtained by winding the wire into a coil.

An **electromagnet** is a coil of wire with a current flowing through it. The magnetic field of the electromagnet is only present when the current is flowing. If the current is switched off, the coil is no longer magnetic.

The strength of an electromagnet can be increased by:

• increasing the current
• increasing the number of turns of wire in the same length of coil
• adding a **core** made of a magnetic material.

A bar magnet is a **permanent** magnet because it is always magnetic. Electromagnets are used where the magnetism needs to be variable, or needs to be switched on and off.

Magnetic field patterns

Straight wire

The magnetic field pattern around a straight wire carrying a current is a series of circles. The direction of the field depends on the direction of the current. The circles are drawn closer together near the wire to show that the field is stronger nearer the wire.

(a)

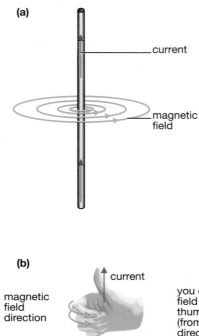

current

magnetic field

(b)

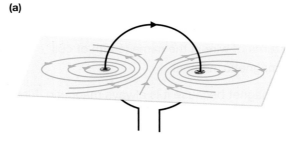

current

magnetic field direction

you can remember the direction of the field using your right hand; point your thumb in the direction of the current (from + to –), and your fingers show the direction of the magnetic field

Fig. 6b.01: (a) The magnetic field pattern around a straight wire carrying a current
(b) Finding the direction of the field

Flat coil

If a wire is made into a flat coil, it produces a magnetic field like the one shown in Fig. 6b.02 when the current is flowing. The magnetic field is strongest in the centre of the coil.

(a)

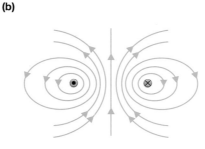

(b)

Fig. 6b.02: (a) The magnetic field around a flat coil of wire carrying a current
(b) Plan view of the magnetic field; a dot is used to show the current coming out of the paper, and a cross shows current going into the paper

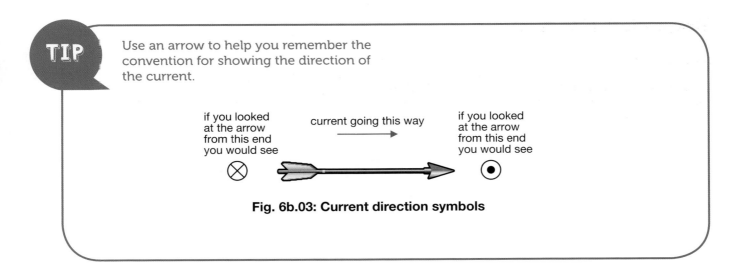

TIP

Use an arrow to help you remember the convention for showing the direction of the current.

if you looked at the arrow from this end you would see

current going this way

if you looked at the arrow from this end you would see

Fig. 6b.03: Current direction symbols

Solenoid

The magnetic field around a long coil of wire is shown in Fig. 6b.04. A coil of wire like this is called a **solenoid**.

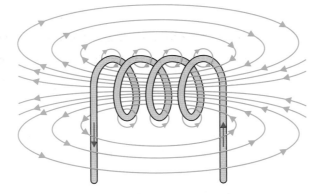

Fig. 6b.04: Magnetic field around a solenoid

CAM

The magnetic field is strongest inside the coil. The strength of the magnetic field can be changed by changing the size of the current. The direction of the magnetic field can be changed by changing the direction of the current.

The motor effect

A charged particle moving across a magnetic field will experience a force, as long as it is not moving parallel to the magnetic field lines. A current in a wire is a flow of charged particles, so a wire carrying a current in a magnetic field will experience a force.

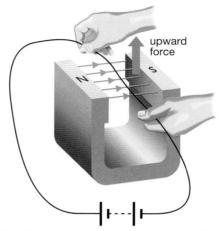

Fig. 6b.05: Apparatus to demonstrate the motor effect

You can predict the direction of the force using **Fleming's left hand rule** (Fig. 6b.06), using a current direction that goes from + to –.

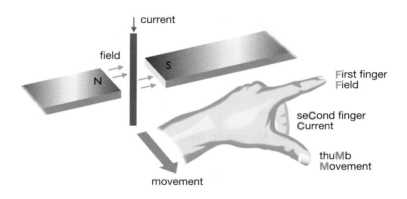
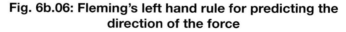

Fig. 6b.06: Fleming's left hand rule for predicting the direction of the force

> **TIP**
>
> Electrons in a wire flow from the negative to the positive terminal of a cell. However, electrons had not been discovered when scientists first started to investigate the magnetic effects of current. They needed to agree on a 'direction' for the current, and decided that they would talk about it as if it flowed from the positive to the negative terminal.
>
> This is why 'conventional current' is from + to –, and it is this conventional current direction that is used when we are thinking about electromagnets, motors and generators.

The size of the force produced increases if the:

• size of the current increases
• strength of the magnetic field increases.

The direction of the force reverses if the:

• direction of the current reverses
• direction of the magnetic field reverses.

Electric motors

Fig. 6b.07 shows how the motor effect is used in a simple electric motor. The magnets provide a uniform magnetic field, and a flat coil is mounted so that it can spin between the magnets. Contact is made with an electric circuit by a split metal ring and **carbon brushes**. The **split ring commutator** swaps the connections over every half turn, so that the coil keeps spinning in the same direction. This kind of motor needs a **direct current**.

The force provided by the motor can be increased by:

• increasing the current
• increasing the strength of the magnetic field
• increasing the number of turns on the coil.

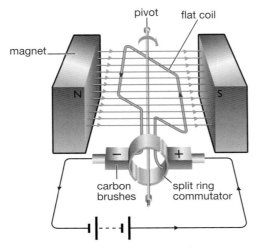

Fig. 6b.07: A simple electric motor

TIP

Look at Fig. 6b.07 and use the left hand rule to work out which side of the coil should be moving up and which side should be moving down. Is the arrow showing the correct direction of rotation?

When the coil is vertical there is no force making it spin, because the sides of the coil are now moving along the field lines. The momentum of the spinning coil makes it carry on turning.

Loudspeakers

Loudspeakers make use of the motor effect. A loudspeaker uses a permanent magnet shaped as shown in Fig. 6b.08. There is a uniform magnetic field between the central north pole of the magnet and the surrounding south pole.

When a current flows through the coil the motor effect makes the coil move. The loudspeaker cone is attached to the coil, so that also moves. The current to the coil changes continually, making the loudspeaker vibrate.

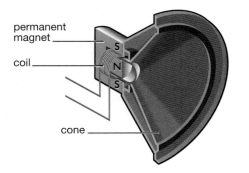

Fig. 6b.08: A loudspeaker

CAM

Relays

Electromagnets are used in **relays**. Relays are used to allow a small current to switch on a separate circuit that may be carrying a much greater current.

The electromagnet is arranged so that it attracts a piece of iron when a current flows through it. When the piece of iron moves, it closes the contacts in the power circuit.

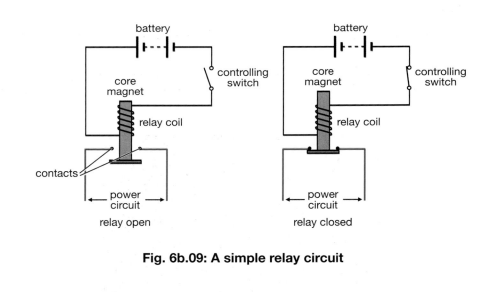

Fig. 6b.09: A simple relay circuit

Cathode rays

Cathode rays are beams of electrons. They are made in **cathode ray tubes**.

A cathode ray tube is a glass vessel with a vacuum inside, and a pair of electrodes.

- One electrode is a filament that is heated by an electric current. This filament gets hot enough to knock electrons off the atoms near the surface of the filament. This emission of electrons due to heating is called **thermionic emission**.
- The heated filament is connected to the negative end of a cell or power supply.
- The other electrode is connected to the positive end.
- The negative charge on the filament repels the electrons and the positive charge attracts them, so they travel in straight lines through the tube.
- The negative electrode is also called the **cathode**, so the beams of electrons are called **cathode rays**. The positive electrode is the **anode**.
- The cathode rays are detected because they cause fluorescent screens to glow.

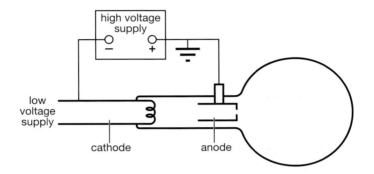

Fig. 6b.10: A cathode ray tube

Deflecting cathode rays

If a pair of magnets is held either side of the tube, the cathode rays will bend because they are charged particles moving through a magnetic field.

Fig. 6b.11 shows a version of a cathode ray tube designed to show the effect of an electric field on cathode rays. A voltage is applied across the top and bottom plates. The cathode rays are negative charges, so they are deflected towards the positive plate.

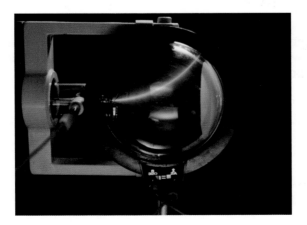

Fig. 6b.11: Deflecting cathode rays with an electric field

Oscilloscopes

An oscilloscope is based on a cathode ray tube. It has an electron gun that produces a beam of electrons. It also has two sets of plates that have varying voltages applied to them. One set of plates deflects the beam up and down, and the other set deflects it from side to side.

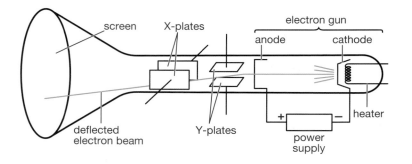

Fig. 6b.12: The parts of a cathode ray oscilloscope

Fig. 6b.13: A cathode ray oscilloscope

A **cathode ray oscilloscope** is used to display waveforms, such as sound waves. The X-plates make the beam sweep from side to side. The speed of the beam from side to side can be adjusted to give different time scales on the oscilloscope screen.

To display the waveform of a high frequency sound, the beam would need to sweep across the screen very fast.

If a sound wave is being displayed, the signals from a microphone are connected to the Y-plates, and deflect the beam up and down. The combination of the movement up and down with the movement of the beam across the screen produces a waveform on the screen.

You should now be able to:

* state that an electric current in a conductor produces a magnetic field around it (see page 153)
* describe an electromagnet (see page 153)
* describe the factors that affect the strength of an electromagnet (see page 153)
* describe the magnetic field patterns for a straight wire, a circular coil and a solenoid (see page 154)
* state that there is a force on a charged particle moving across a magnetic field (see page 156)
* use the left hand rule to predict the direction of the force on a wire carrying a current in a magnetic field (see page 156)
* explain how the force on a wire carrying a current in a magnetic field is used in motors and loudspeakers (see pages 157 and 158)
* state the factors that affect the size of the force on a wire carrying a current in a magnetic field (see pages 157 and 158)

CAM * describe how a relay works (see page 158)
* describe the production and detection of cathode rays, and how they are affected by magnetic and electric fields (see page 159)
* describe how a cathode ray oscilloscope works (see page 160).

Review questions

1. Describe three ways of reducing the strength of an electromagnet.

2. Sketch the magnetic field pattern for:

    (a) a wire carrying a current

    (b) a flat coil carrying a current

    (c) a solenoid.

3. What do your two fingers and thumb represent when you are using the left hand rule?

4. Give two ways of increasing the force produced by the motor effect.

5. What do the following parts of a simple electric motor do?

    (a) two magnets

    (b) carbon brushes

    (c) split ring commutator.

CAM 6. Briefly explain how a relay works.

7. In a cathode ray oscilloscope:

    (a) what do the X-plates do?

    (b) what do the Y-plates do?

Practice questions

1. The diagram shows an electric bell. When it is switched on, current flows around the circuit.

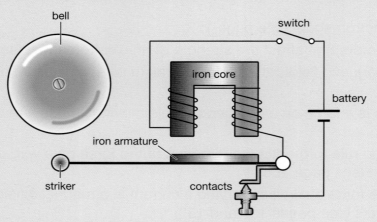

(a) Sketch the shape of the magnetic field around a straight wire. **(2)**

(b) A wire wrapped around a core forms an electromagnet. Write down three ways of increasing the strength of an electromagnet. **(3)**

(c) How does the bell work? Explain what happens when the switch is closed. **(6)**

(d) Should the iron core be made from a magnetically hard or magnetically soft form of iron? Explain your answer. **(2)**

2. The diagram shows a device called Barlow's wheel. When a cell is connected to the wheel and to the bath of mercury, as shown, the disc spins.

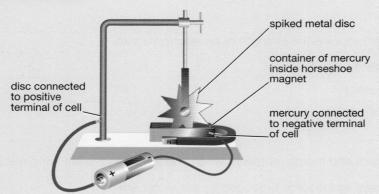

(a) Which way will the disc spin? Explain your reasoning. **(4)**

(b) Describe two ways in which you could make the disc spin faster. **(2)**

(c) Describe two ways in which you could make the disc spin in the other direction. **(2)**

CAM 3. The diagram shows a cathode ray oscilloscope.

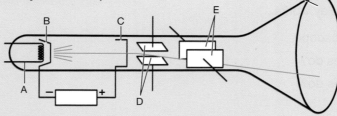

Name parts A to F and describe their functions. **(6)**

C Electromagnetic induction

You will be expected to:

CAM marks in the margin indicate the following points.

* ★ recall that a voltage is induced when a conductor is in a changing magnetic field
* ★ recall the factors that affect the size of the induced voltage
* **CAM** ★ understand that the direction of an induced e.m.f. opposes the change causing it
* ★ describe how a generator works, and the factors which affect the size of the induced voltage
* **CAM** ★ sketch a graph of voltage against time for a simple a.c. generator
* ★ describe the structure of a transformer
* ★ recall how the number of turns on the coils in a transformer affects the output voltage
* **CAM** ★ describe how transformers work
* ★ explain how step-up and step-down transformers are used in transmitting electricity
* ★ recall and use the formula relating voltage and number of turns in the primary and secondary coils of a transformer
* ★ recall and use the formula relating input and output power of a transformer.

Units

You will use the following quantities and units in this section.

Quantity	Symbol	Unit
current	I	amperes (or amps) (A)
voltage	V	volts (V)
number of turns of wire in a coil	n	no units

Inducing a voltage

If a wire is moved through a magnetic field, a voltage is **induced** in the wire. If the wire is part of a circuit, the induced voltage will make a current flow (Fig. 6c.01).

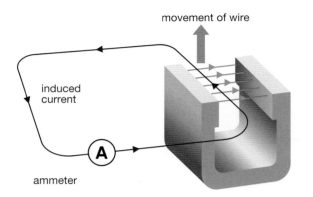

Fig. 6c.01: Electromagnetic induction

The same effect occurs if a magnetic field changes around the wire. The changing magnetic field can be from:

- a moving magnet
- a spinning magnet
- an electromagnet being turned on and off.

The size of the induced voltage can be increased by:

- increasing the strength of the magnetic field
- increasing the speed of movement of the magnet or the coil of wire
- increasing the rate at which the magnetic field changes.

CAM

The direction of the induced e.m.f. opposes the change causing it. If you apply the left hand rule (page 156) to the current and magnetic field shown in Fig. 6c.01, you should find that the motor effect will produce a force acting downward, which will oppose the upward movement of the wire.

Generating electricity

Electricity can be generated by spinning a magnet inside a coil of wire, or by spinning a coil of wire inside a magnetic field (Fig. 6c.02).

The **slip rings** and **carbon brushes** provide an electrical contact between the spinning coil and the external circuit.

This kind of generator produces **alternating current**. In an alternating current, the electrons change direction many times each second.

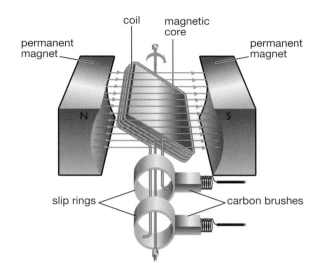

Fig. 6c.02: A simple generator

The size of the induced voltage can be increased by:

- increasing the strength of the magnetic field
- increasing the rate at which the coil spins
- increasing the number of turns of wire on the coil
- using a coil with a core made of a magnetic material.

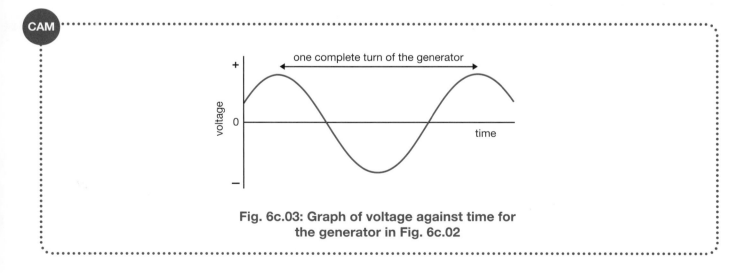

Fig. 6c.03: Graph of voltage against time for the generator in Fig. 6c.02

Transformers

A **transformer** is used to change an alternating voltage.

A transformer consists of an **iron core** with two **coils** of wire wound on it. The voltage to be changed is applied to the **input coil** (also called the **primary coil**). The **output coil** (also called the **secondary coil**) provides the changed voltage.

The output voltage depends on the input voltage and on the number of turns in the primary and secondary coils.

- If the secondary coil has *more* turns than the primary coil, the output voltage is *higher* than the input voltage, and the transformer is a **step-up transformer**.
- If the secondary coil has *fewer* turns than the primary coil, the output voltage is *lower* than the input voltage, and the transformer is a **step-down transformer**.

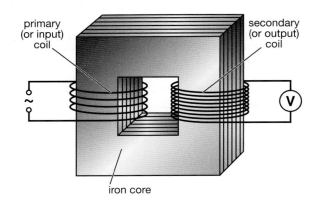

Fig. 6c.04: The structure of a transformer: this is a step-up transformer

CAM

How transformers work

When the primary coil is connected to an a.c. electricity supply, it produces a magnetic field that is continuously changing. The iron core concentrates this magnetic field. As the secondary coil is wound onto the same iron core, the changing magnetic field induces an e.m.f. in the coil.

TIP Remember that transformers only work with *alternating* current.

Transformers and electricity transmission

Electricity is sent from power stations to customers by **transmission lines**. Transmission lines are just long wires, and waste some energy by heating in the same way as any other wires. The amount of energy wasted depends on the current: less energy is wasted at a lower current.

The power of an electric current depends on the current and the voltage (see page 46). Energy is conserved, so if the voltage is increased the current decreases.

Before electricity is sent from power stations, a step-up transformer is used to increase the voltage of the electricity produced. This reduces the current and so reduces the energy losses in the transmission lines. Step-down transformers are used in factories, or in electricity substations in towns, to reduce the voltage to safer levels before the electricity is used. In many countries, the voltage for houses, shops and offices is 230 V. Other countries use voltages of 110 V or 120 V.

Calculating the input or output voltage

The ratio of the input and output voltages in a transformer depends on the ratio of the number of turns in the two coils:

$$\frac{\text{voltage in primary coil}}{\text{voltage in secondary coil}} = \frac{\text{number of turns in primary coil}}{\text{number of turns in secondary coil}}$$

$$\frac{V_p}{V_s} = \frac{n_p}{n_s}$$

The formula can be rearranged into the following versions:

$$V_p = \frac{V_s \times n_p}{n_s}$$

$$V_s = \frac{V_p \times n_s}{n_p}$$

$$n_p = \frac{n_s \times V_p}{V_s}$$

$$n_s = \frac{n_p \times V_s}{V_p}$$

TIP

You need to remember this formula, as it will not be given to you in an exam. You also need to be able to rearrange it.

You can learn the different versions of the formula given here, but it is better if you learn the main formula and know how to rearrange formulae algebraically (see Appendix 1).

Worked example

A transformer has 150 turns on the primary coil and 20 turns on the secondary coil. If the input voltage is 230 V, what is the output voltage?

Answer

$$V_s = \frac{V_p \times n_s}{n_p}$$

$$= \frac{230 \text{ V} \times 20}{150}$$

$$= 30.7 \text{ V}$$

TIP

Check your answer for sense after carrying out transformer calculations. If the number of turns in the primary coil is bigger than the number in the secondary coil, then the primary (or input) voltage must be bigger than the secondary (or output) voltage.

Transformers and power

Energy cannot be created or destroyed, so the energy transferred to a transformer each second must be the same as the energy transferred from it:

$$\text{input power} = \text{output power}$$

The power transferred by an electric current is the voltage multiplied by the current (page 46), so if the transformer is 100% efficient:

$$V_p \times I_p = V_s \times I_s$$

The formula can be rearranged into the following versions:

$$V_p = \frac{V_s \times I_s}{I_p}$$

$$V_s = \frac{V_p \times I_p}{I_s}$$

$$I_p = \frac{V_s \times I_s}{V_p}$$

$$I_s = \frac{V_p \times I_p}{V_s}$$

> **TIP** The formula will be given to you in the exam but you do need to be able to rearrange it (or to learn all the above versions).

Worked example

A transformer converts the 230 V mains supply to provide a 12 V, 3 A supply. What is the current in the primary coil?

Answer

$$I_p = \frac{V_s \times I_s}{V_p}$$

$$= \frac{12 \text{ V} \times 3 \text{ A}}{230 \text{ V}}$$

$$= 0.16 \text{ A}$$

> **TIP** Check that your answer is sensible. If the secondary current is higher than the primary current, the secondary voltage must be *lower* than the primary voltage.

You should now be able to:

- ★ describe how to use a magnetic field to induce a voltage in a wire (see page 164)
- ★ list the factors that affect the size of the induced voltage (see page 164)
- **CAM** ★ understand that the direction of an induced e.m.f. opposes the change causing it (see page 164)
- ★ describe how a generator works (see page165)
- ★ describe the factors which affect the size of the voltage induced by a generator (see page 165)
- **CAM** ★ sketch a graph of voltage against time for a simple a.c. generator (see page 165)
- ★ describe the structure of a transformer (see page 166)
- ★ state how the numbers of turns on the coils in a transformer affects the output voltage (see page 166)
- **CAM** ★ describe how transformers work (see page 166)
- ★ explain how step-up and step-down transformers are used in transmitting electricity (see page 166)
- ★ state and use the formula relating voltage and number of turns in the primary and secondary coils of a transformer (see page 167)
- ★ state and use the formula relating input and output power of a transformer (see page 168).

Review questions

1. A transformer has 50 turns on the primary coil and 400 turns on the secondary coil. If the secondary voltage is 50 V, what is the primary voltage?

2. The input of a transformer is a current of 20 A at 120 V. If the output current is 0.5 A, what is the output voltage?

Practice questions

1. The diagram shows a permanent magnet being moved into a coil of wire.

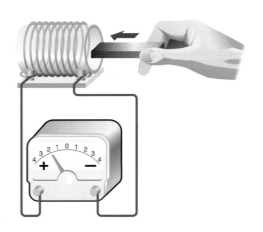

   (a) What will the ammeter read when the magnet is fully inside the coil? Explain your answer. (2)

   (b) What will the ammeter read when the magnet is being removed from the coil? Explain your answer. (2)

   (c) Describe three changes that could be made to increase the size of the induced current. (3)

   (d) List two differences between this apparatus and a simple generator. (2)

2. Transformers are used in the electricity grids that are used to distribute electricity from power stations to consumers.

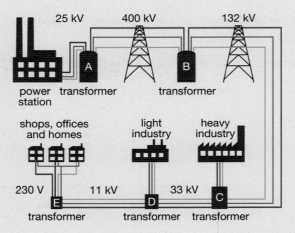

   (a) What do transformers do? (2)

   (b) Which transformers in the diagram are step-down transformers? (1)

   (c) If transformer A has 1000 turns on its primary coil, how many does it have on its secondary coil? (3)

   (d) The power station output is 2 gigawatts (2×10^9 W).

   (i) How large is the power transmitted by the 400 kV line compared to the power supplied from the power station? (2)

   (ii) How large is the current in the 400 kV line compared to the current from the power station? Explain your answer. (4)

Section Seven

7 Radioactivity and particles

A Radioactivity

You will be expected to:

★ describe the structure of an atom
★ use symbols to show mass number and atomic number
★ explain what isotopes are
★ describe the nature and properties of alpha particles, beta particles and gamma radiation
★ describe how the emission of each type of radiation affects the mass number and atomic number of an atom
★ balance nuclear equations
★ explain how ionising radiations can be detected
★ recall the sources of background radiation
★ describe how the activity of a source decreases with time
★ describe the meaning of the term half-life
★ use the idea of half-life in calculations
★ describe some uses of radioactivity
★ describe some dangers of radioactivity and how to reduce the risks.

Units

You will use the following quantities and units in this section.

Quantity	Symbol	Unit
activity		becquerels (Bq)
length		centimetres (cm)
time	t	hours (h), minutes (min), seconds (s)

Atomic structure

An atom consists of protons, neutrons and electrons, as summarised in the table below.

Particle	Location	Mass	Charge
proton	nucleus	1	+1
neutron	nucleus	1	0
electron	around the nucleus	negligible	−1

- The number of protons in every atom of an element is the **atomic number** (or **proton number**) of that element.
- All atoms of a particular element *always* have the same number of protons. For example, all carbon atoms have 6 protons in their nucleus.
- The total number of particles in the nucleus of an atom is the **mass number** (or **nucleon number**).

> **TIP**
> Remember that the mass number is the *total* number of particles in the nucleus, not the number of neutrons.

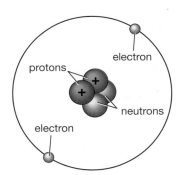

Fig. 7a.01: Particles in an atom (not to scale)

CAM

A **nuclide** (a particular nucleus) can be represented by the symbol for the element, the proton number Z and the nucleon number A:

$$^{A}_{Z}X$$

Any nucleus can be described using the symbol for the element and the atomic and mass numbers. For example, carbon can be represented like this:

$$^{12}_{6}C$$

Worked example

A lithium atom (Li) has 3 protons in its nucleus, and 4 neutrons. Describe a lithium nucleus using a symbol.

Answer

Atomic number = 3
Mass number = 3 + 4 = 7

$$^{7}_{3}Li$$

Isotopes

All atoms of a particular element have the same number of protons, but they can have different numbers of neutrons.

Atoms of the same element with different numbers of neutrons are called **isotopes**. For example, most carbon atoms have 6 neutrons, but some have 8. The mass number of this isotope of carbon is 14, so it is called carbon-14.

Radiation

Some isotopes are **unstable**, and can emit radiation in the form of alpha particles, beta particles or gamma rays. These are all forms of **ionising radiation**, which means that the radiation can knock electrons off atoms, leaving charged **ions** behind.

Alpha particles

Alpha (α) particles consist of two protons and two neutrons.

This is the same composition as a helium nucleus, so an alpha particle is represented like this:

$$^{4}_{2}\text{He}$$

An alpha (α) particle has a +2 charge, because it has 2 protons and no electrons.

Energy carried by an alpha particle is transferred quickly to any material it enters, because the particle is large and massive. This means that α-radiation:

- is highly ionising
- does not **penetrate** far.

Beta particles

Beta (β) particles are fast-moving electrons that come from the nucleus of atoms when a neutron decays to form a β beta particle and a proton. A particle is represented like this:

$$^{0}_{-1}\text{e}$$

A beta particle has a –1 charge. It is much less massive than an alpha particle, so it:

- is not as ionising as an alpha particle
- penetrates further through materials.

Gamma rays

> **TIP**
> Remember that a beta particle is an electron *from the nucleus* of an atom. It is not one of the electrons that normally move around the outside of the nucleus.

Gamma (γ) rays are electromagnetic waves with very short wavelengths. They are not particles. They do not interact much with atoms in matter, so gamma rays:

- are only weakly ionising
- penetrate a long way.

Comparing radiation

The penetrating power of different forms of radiation can be investigated using the apparatus shown in Fig. 7a.02.

- A source emitting one type of radiation is put in the apparatus.
- Different materials can be placed between the source and the detector, or different thicknesses of the same material can be used.
- The amount of radiation passing through the materials is measured by the detector.

Atomic changes and nuclear equations

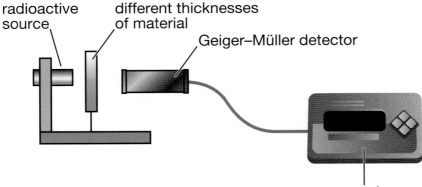

Fig. 7a.02: Comparing the penetrating power of alpha, beta and gamma radiation

Radiation	Penetrating power	Radiation can be stopped by	Ionising power
alpha (α)	weak	a few cm of air human skin paper	strong
beta (β)	medium	1 metre of air thin aluminium	medium
gamma (γ)	strong	thick lead sheet	weak

Electric and magnetic fields

Charged particles are deflected when they move through electric or magnetic fields. Fig. 7a.03 shows how alpha and beta particles are detected. Gamma rays are not charged particles so they are not deflected.

Alpha and beta particles are deflected in opposite directions because they have opposite charges. There is a greater force on alpha particles than on beta particles because they have twice as much charge. However, the beta particles are deflected more than the alpha particles because their mass is much, much smaller than the mass of alpha particles.

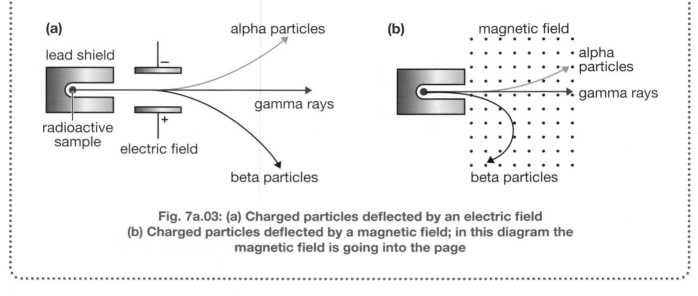

Fig. 7a.03: (a) Charged particles deflected by an electric field
(b) Charged particles deflected by a magnetic field; in this diagram the magnetic field is going into the page

Emission of an alpha particle

When an alpha particle is emitted by an unstable nucleus, the nucleus has lost two protons and two neutrons. Its atomic number goes *down* by 2 and its mass number goes *down* by 4.

Since the number of protons has changed, the nucleus is now a different element. This new element is called a **daughter isotope**.

Emission of a beta particle

When a beta particle is emitted by an unstable nucleus, a neutron in the nucleus has been converted to a proton, and an electron is emitted. The atomic number of the nucleus goes *up* by 1 and its mass number *does not change*.

Emission of a beta particle also changes the element.

Emission of a gamma ray

Emission of a gamma ray removes some energy from an unstable nucleus, but it does not change the atomic number or mass number. The element does not change.

The change caused by a radioactive decay can be represented by a nuclear equation. In a similar way to chemical equations, a nuclear equation must be balanced, with the same number of protons and neutrons on each side of the equation.

Worked examples

Example 1

A radium (Ra) isotope has an atomic number of 88 and a mass number of 222. Write a nuclear equation to show what happens when it undergoes alpha decay.

Answer

Alpha decay will decrease the atomic number by 2. The element with an atomic number of 86 is radon (Rn). The radon atom formed will have a mass number of 218.

$$^{222}_{88}\text{Ra} \rightarrow \ ^{218}_{86}\text{Rn} + \ ^{4}_{2}\text{He}$$

Example 2

Carbon-14 undergoes beta decay to form nitrogen. Write a nuclear equation to show this.

Answer

Beta decay increases the atomic number by 1 and does not change the mass number.

$$^{14}_{6}\text{C} \rightarrow \ ^{14}_{7}\text{N} + \ ^{0}_{-1}\text{e}$$

 TIP

Check that your nuclear equation is balanced by adding up the mass numbers on the right-hand side of the equation. The total should be the same as the mass number of the starting isotope on the left-hand side.

Do the same for the atomic numbers.

Detecting radiation

Ionising radiation can be detected by using:

- a Geiger–Müller detector that reacts to each particle of ionising radiation that enters it
- photographic film – energy carried by the radiation darkens the film.

People who work with radioactive materials often have to wear film badges. The film is removed from the badge periodically and developed to show the amount of ionising radiation to which the person has been exposed.

Fig. 7a.04: A film badge records the amount of each type of radiation a person has received

Background radiation

Radiation is being produced around us all the time. This is called **background radiation**.

Sources of background radiation include:

- radon gas, which is a gas emitted by radioactive elements in certain types of rock
- cosmic rays – radiation from space
- internal radiation, from atoms within our own bodies
- food and drink
- buildings and the ground
- radioactive materials used in medicine
- fallout from nuclear tests and accidents at nuclear power stations.

The amount that each of these sources contributes to the background radiation varies from place to place.

Activity and half-life

The **activity** of a radioactive source is the number of atoms that decay each second, and is measured in becquerels (Bq).

The decay of radioactive isotopes is a random process. It is not possible to predict when any one atom will decay. However, there is an average rate of decay for each radioactive element, and this is measured by the **half-life**.

The more atoms of a radioactive element there are in a sample, the more will decay each second. As some of the atoms in the sample decay (and therefore change into different elements), there are fewer radioactive atoms left, and so the activity decreases.

The half-life is the length of time it takes for half of the radioactive atoms in a sample to decay. This is also the time it takes for the activity of a sample to halve.

You can work out the half-life of an isotope from a graph of activity against time (Fig. 7a.05):

- take any activity (such as 160 Bq at the start of the graph), and work out half its value (80 Bq)
- read off the times for these two points (0 and 2 minutes – the red dashed line on Fig. 7a.05
- this time gives you the half life.

This works for any initial activity, as shown on the graph by the green lines.

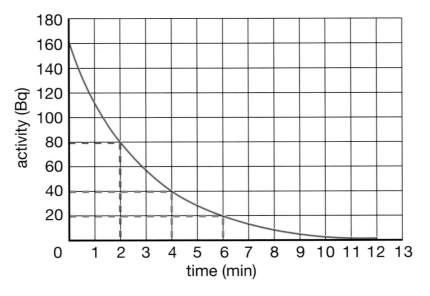

Fig. 7a.05: Working out the half-life from a graph of activity against time

You may also be asked to work out how long it will take for activity to decay to a certain value. There is no equation for this. You usually just need to add up the half-lives, as shown in the examples below.

Worked examples

Example 1

A radioactive isotope has a half-life of 3 days. How long will it take for the activity to decrease from 1000 Bq to 125 Bq?

Answer

After 1 half-life the activity will be 500 Bq.
After 2 half-lives the activity will be 250 Bq.
After 3 half-lives the activity will be 125 Bq.

It will take 3 × 3 days = 9 days for the activity to decrease from 1000 Bq to 125 Bq.

Example 2

A radioactive isotope has a half-life of 15 minutes. Its activity is 800 Bq. What will its activity be after 1 hour?

Answer

1 hour is 4 half-lives.

After 1 half-life the activity will be 400 Bq.
After 2 half-lives the activity will be 200 Bq.
After 3 half-lives the activity will be 100 Bq.
After 4 half-lives the activity will be 50 Bq.

After 1 hour the activity will be 50 Bq.

Radioactivity in medicine

Radioactive isotopes are used as **tracers**. In medicine, a patient is given a dose of a compound containing radioactive isotopes that will emit gamma rays. A gamma ray camera is used to follow the progress of the radioisotope around the body. The compound used depends on what is being investigated.

Radioactive isotopes are also used in **radiotherapy**, such as to treat cancer. Radioactive isotopes might be given to a patient as part of a compound that will collect near cancer cells, for example. Alpha or beta radiation emitted by these isotopes will kill nearby cells, including the cancer cells.

Sources of gamma radiation are used in medicine to sterilise surgical instruments.

Non-medical applications

Radioactive isotopes are also used as tracers in non-medical applications.

- A small amount of a radioactive material might be added to water in a pipe where a leak was suspected. The radioactive isotopes will leak out of the pipe along with the water, and can be detected using a Geiger–Müller detector.

Radioactivity can also be used to date rocks or archaeological specimens. The best known of these methods is **radiocarbon dating**.

- The atmosphere contains a constant proportion of carbon-14, because it is being formed by cosmic rays in the upper atmosphere at about the same rate as it decays.
- Living organisms take in carbon in photosynthesis (in plants) or in their food, and so living plants and animals contain a steady proportion of carbon-14.
- When organisms die, the carbon-14 in their bodies continues to decay but no more is being absorbed, so the proportion of carbon-14 goes down.
- Scientists can find out how long ago an animal or person died (or how long ago a tree was chopped down to make part of a building) by measuring the proportion of carbon-14 in it.

A similar technique can be used to find the age of certain rocks. If a rock contains a radioactive isotope, its age can be worked out by comparing the amount of that isotope it contains with the amount of daughter isotopes present.

Dangers of radioactivity

Radiation can damage cells and tissues in the body directly, causing burns. It can also cause **mutations** in living organisms. These mutations may cause diseases such as cancer, or may cause birth defects if the person later has children.

Anyone who uses radioactive materials in their job must know how to use them safely. Precautions for using radioactive materials in schools include:

- keeping radioactive sources in lead-lined boxes
- storing them in securely locked cupboards
- only handling the sources with tongs (to keep hands as far away from the source as possible)
- never pointing a source directly at anyone.

People who work with radioactive materials all the time wear film badges (see page 178), to monitor the amount of radiation they are exposed to. There are laws that limit the amount of radiation anyone can be exposed to in a year – if a worker exceeds this dose they will have to be given work away from radiation for a while.

Radioactive waste

Radioactive waste is produced by hospitals, industry and scientific laboratories, as well as by the nuclear power industry. Radioactive waste must be disposed of safely, but this is not always easy. Some radioactive isotopes used in the nuclear industry have half-lives of hundreds of thousands of years.

Waste that is not very radioactive, or that has a short half-life, can be stored in a safe place until its activity has reduced to safe levels. It is then put into landfill sites.

Waste that is very radioactive is turned into a glass-like material and stored. The aim is to store it underground eventually, but so far no suitable sites have been identified. Such sites must be geologically stable (where there is no risk of earthquakes), and must not allow any escape of waste to get into water supplies. People who live in the areas near possible sites do not like the idea of radioactive waste being stored nearby.

You should now be able to:

★ use symbols to show mass number and atomic number (see page 173)
★ explain what isotopes are (see page 174)
★ describe the nature and properties of (a) alpha particles, (b) beta particles, (c) gamma radiation (see page 174)
★ describe how the emission of each type of radiation affects the mass number and atomic number of an atom (see page 176)
★ balance nuclear equations (see page 177)
★ explain how ionising radiations can be detected (see page 178)
★ recall the sources of background radiation (see page 178)
★ describe how the activity of a source decreases with time (see page 179)
★ describe the meaning of the term half-life (see page 179)
★ use the idea of half-life in calculations (see page 180)
★ describe some uses of radioactivity (see page 181)
★ describe some dangers of radioactivity (see page 181).

Review questions

1. An isotope of americium (Am) has 95 protons and 146 neutrons. It decays by alpha emission to form neptunium (Np). Write a balanced nuclear equation for this decay.

2. Caesium-137 (Cs) has an atomic number of 55. Write a balanced nuclear equation for its beta decay to form barium (Ba).

3. A radioactive source has an activity of 120 Bq and a half-life of 30 minutes. What will its activity be 90 minutes later?

4. The activity of a source is found to be 120 Bq. Twenty minutes later the activity is 30 Bq. What is the half-life of the source?

Practice questions

1. The symbols show two different kinds of oxygen atom.

$$^{16}_{8}O \qquad\qquad ^{18}_{8}O$$

(a) (i) What is the number at the bottom of the symbol called? **(1)**

    (ii) What does it represent? **(1)**

(b) (i) What is the number at the top of the symbol called? **(1)**

    (ii) What does it represent? **(1)**

(c) What is the name for different versions of the same element like these? **(1)**

(d) (i) Where are electrons found in an atom? **(1)**

    (ii) How many electrons does an atom of oxygen have? Explain your answer. **(2)**

2. (a) Use the notation shown in question 1 to represent:

    (i) an alpha particle **(1)**

    (ii) a beta particle **(1)**

(b) What are beta particles? **(2)**

(c) An isotope of thorium (Th) has an atomic number of 90 and a mass number of 234. It decays by beta emission to form protactinium (Pa). Write a balanced nuclear equation to represent this decay. **(4)**

(d) A different isotope of thorium has only 140 neutrons in its nucleus, and decays by alpha emission to form radium (Ra). Write a balanced nuclear equation to represent this decay. **(4)**

3. Some houses are fitted with fans and pipes that pump air out of the spaces beneath the floor. These systems are fitted in areas where radon gas seeps out of the ground.

fan

pipe connected to space beneath floor

(a) Why is it unsafe to have high levels of radon gas in a building? **(2)**

(b) Name the instrument used to detect ionising radiation. **(1)**

(c) Name two other sources of background radiation. **(2)**

4. Technetium-99m is a radioactive isotope widely used in medicine. It has a half-life of about 6 hours and decays by the emission of gamma rays.

(a) What does 'half-life' mean? **(1)**

(b) Draw a graph to show the activity of a sample of technetium-99m over a period of 24 hours, assuming its initial activity is 80 Bq. **(3)**

(c) Technetium-99m is used as a tracer in medicine. What does this mean and how is it done? **(3)**

(d) Technicians handling technetium-99m must take precautions to protect themselves against exposure to gamma rays.

    (i) What harm could exposure to gamma rays cause? **(1)**

    (ii) Explain why people working with radioactive materials wear film badges. **(2)**

B Particles

You will be expected to:

★ describe the results of Geiger and Marsden's experiments with gold foil and alpha particles
★ describe Rutherford's nuclear model of the atom
★ explain the factors which affect the deflection of alpha particles by a nucleus
★ describe how a U-235 nucleus is split by a neutron and how energy is released
★ recall that the fission of U-235 produces two daughter nuclei and a small number of neutrons
★ explain how a chain reaction happens
★ describe how the chain reaction is controlled in a nuclear power station
CAM ★ recall that the nuclear reactions in the Sun are fusion reactions.

Alpha particles and gold foil

In 1909 Geiger and Marsden carried out an experiment where they fired a stream of alpha particles at a piece of gold foil, using the apparatus shown in Fig. 7b.01.

The lead shielding helps to protect the scientists. It also makes sure that all the alpha particles are travelling in approximately the same direction when they leave the source.

The tracks of the alpha particles were worked out using the zinc sulfide screen. This produces a **scintillation** (a tiny flash of light) every time an alpha particle hits it.

At the time that Geiger and Marsden did their experiment, scientists thought that the atom was a mass of positive charge with small electric charges embedded in it. This was the '**plum pudding**' model of the atom. Geiger and Marsden expected the alpha particles to go straight through the gold foil. However, they found that some of the particles were deflected, and a few actually bounced back towards the source.

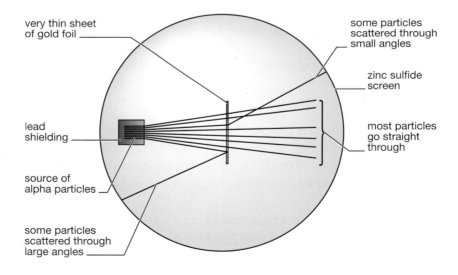

very thin sheet of gold foil

some particles scattered through small angles

zinc sulfide screen

lead shielding

most particles go straight through

source of alpha particles

some particles scattered through large angles

Fig. 7b.01: The Geiger-Marsden experiment and its results

Rutherford's model of the atom

Rutherford used the results of Geiger and Marden's experiment to suggest the **nuclear model** of the atom (which is the model we still use today).

In this model the mass and positive charge in an atom are concentrated in a very small nucleus. The nucleus is much, much smaller than the overall size of the atom. The negative charges (electrons) are spread around the rest of the atom.

This mode explains the results of the experiment:

- the alpha particles that went straight through the gold foil passed through gaps between the nuclei
- the alpha particles that were scattered by small angles passed near enough to a nucleus for the positive charge on the nucleus to repel the positive charge on the alpha particle
- the alpha particles that were scattered through very large angles must have come very close to a nucleus in the gold foil.

The amount of deflection of an alpha particle by a nucleus depends on:

- the speed of the alpha particle (the faster it is moving, the less a given force will deflect its path)
- how close the alpha particle is to the nucleus when it passes (the closer the nucleus, the larger the force and so the larger the deflection)
- the charge on the nucleus (the greater the charge, the greater the force of repulsion between the nucleus and an alpha particle). This only changes if the material the alpha particles are being fired at is changed.

Fission of uranium-235

Some large nuclei can be split up into smaller nuclei if they are hit by a fast-moving neutron. This process is called **fission**.

The fission of **uranium-235** produces two smaller **daughter nuclei** and also produces several more fast-moving neutrons. Fission releases energy in the form of kinetic energy in these **fission products**.

If the fast-moving neutrons produced by the fission of one U-235 atom hit other U-235 nuclei they can cause further fiss on reactions. Neutrons produced in these reactions can cause further fission reactions, and so on. This process is called a **chain reaction**.

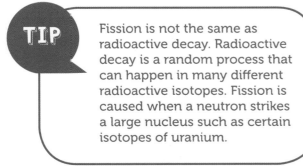

TIP

Fission is not the same as radioactive decay. Radioactive decay is a random process that can happen in many different radioactive isotopes. Fission is caused when a neutron strikes a large nucleus such as certain isotopes of uranium.

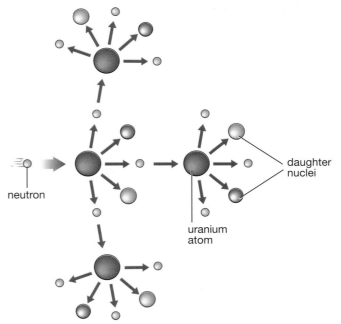

neutron

daughter nuclei

uranium atom

Fig. 7b.02: A chain reaction

Nuclear power stations

Nuclear power stations use a *controlled* chain reaction to produce heat. The heat energy is used to turn water into steam, and the steam turns turbines in the same way as steam produced in fossil fuel power stations does.

Nuclear fuel is contained in **fuel rods** which slot into the reactor **core**.

The core contains a **moderator**, which can be water or graphite. The job of the moderator is to slow down the neutrons. Without the moderator, most of the neutrons released would be moving too fast to cause fission reactions when they passed through the fuel rods.

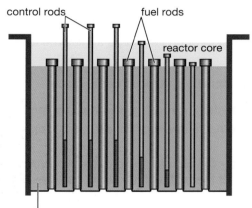

Fig. 7b.03: The core of a nuclear reactor

The reactor also contains **control rods** which can be moved up and down. The control rods absorb neutrons. If the control rods are inserted fully into the core they will absorb most of the neutrons released when U-235 atoms split up, and so the chain reaction will slow down and stop. If the control rods are pulled out, more of the neutrons can cause further fission reactions, so the chain reaction speeds up.

CAM

Nuclear reactions in the Sun

Energy from the Sun also comes from nuclear reactions, but these are **fusion** reactions. In a fusion reaction, two small nuclei fuse together to form a larger nucleus. Most of the energy in the Sun is released as hydrogen nuclei fuse to form helium nuclei, but energy is also released when larger nuclei (up to the mass of iron nuclei) fuse together.

TIP Remember that in fission reactions, large nuclei are split up to form smaller ones. In fusion, small nuclei join together to make bigger ones.

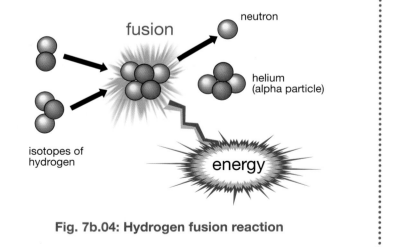

Fig. 7b.04: Hydrogen fusion reaction

You should now be able to:

★ describe Geiger and Marsden's experiments with gold foil and alpha particles and describe what they found (see page 184)

★ describe Rutherford's nuclear model of the atom and how it explains the results of Geiger and Marsden's experiment (see page 185)

★ explain the factors which affect the deflection of alpha particles by a nucleus (see page 185)

★ describe what causes a U-235 nucleus to undergo fission and how energy is released (see page 185)

★ recall the fission products of U-235 (see page 185)

★ explain how a chain reaction happens (see page 185)

★ describe how the chain reaction is controlled in a nuclear power station (see page186)

CAM ★ recall that the nuclear reactions in the Sun are fusion reactions (see page 186).

Review questions

1. (a) What did Geiger and Marsden expect to happen in their experiment?

   (b) Why did they expect this?

2. (a) What were the results of Geiger and Marsden's experiment?

   (b) How does the Rutherford model of the atom explain their results?

3. Explain how the following changes will affect the deflection of an alpha particle.

   (a) The alpha particle moves more slowly.

   (b) The foil is changed to a material with a lower atomic number.

4. What do the following parts of a nuclear reactor do?

   (a) moderator

   (b) control rods

Practice questions

1. The apparatus shown below was used by Geiger and Marsden to investigate the structure of atoms. They fired alpha particles at a very thin sheet of gold foil.

    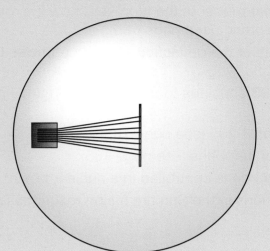

    (a) Describe the results of their investigation. **(3)**

    (b) Sketch the 'Rutherford' model of the atom that was worked out partly as a result of these experiments. **(3)**

    (c) Explain how the results of Geiger and Marsden's experiment support the Rutherford model of the atom. **(3)**

2. The first nuclear weapon used in war consisted of two pieces of uranium. Each piece on its own was too small to sustain a chain reaction. When the bomb was detonated a normal explosive was used to push the two pieces together. Together, the two pieces exceeded the 'critical mass' needed for a chain reaction to take place, and the bomb exploded.

    (a) What are the fission products of uranium-235? **(2)**

    (b) What happens in a chain reaction? **(3)**

    (c) Suggest why a chain reaction does not occur in a small piece of uranium. **(1)**

    (d) A nuclear bomb is an uncontrolled chain reaction. How is the chain reaction controlled in a nuclear power station to provide a continuous supply of heat energy without causing an explosion? **(4)**

Section Eight

The Edexcel International GCSE Physics specification includes some statements that say that you should be able to carry out particular practical activities or investigations. It also has a set of assessment objectives that list some of the practical skills you should develop during the course. These skills will be tested by questions in the two examination papers that you will sit at the end of the course.

CAM

If you are taking the CIE exams, one of your three exam papers will test your practical skills. Your school can choose one of three ways to do this:

- Paper 4 – assess you during the whole course
- Paper 5 – enter you for the practical exam
- Paper 6 – enter you for the written paper.

Whichever method your school has chosen for assessing you, you will be tested on the same set of skills.

The particular practical tasks that you are expected to know how to carry out are covered in the relevant places in Sections 1 to 7. They are listed below and on the following page, just as a reminder. Although some of these statements are only applicable to one of the two specifications covered by this book, it is worth looking at all of them, as questions in the exam may ask you to *apply* your knowledge to new situations.

An investigation involves you finding out how one variable depends on another, such as finding out how current varies with voltage in different components. A practical task is a way of showing or measuring something, such as measuring the speed of sound in air, or showing the shape of a magnetic field.

Specification statement	Practical task or investigation	Page
Edexcel 2.11	investigate how current varies with voltage in different components	56
Edexcel 3.17	investigate the refraction of light using rectangular blocks, semicircular blocks and triangular prisms	88
Edexcel 3.19	determine the refractive index of glass, using a glass block	88
Edexcel 3.28	measure the speed of sound in air	95
Edexcel 3.30	use an oscilloscope to find the frequency of a sound wave	95
Edexcel 5.3	determine density from measurements of mass and volume	126

Specification statement	Practical task or investigation	Page
CIE 1.4	determine density from measurements of mass and volume	26
CIE 1.5(a)	describe how to obtain data to plot extension/load graphs for a spring or wire	29
CIE 1.5(b)	perform and describe an experiment to show that there is no net moment on a body in equilibrium	27
CIE 2.2(c)	describe an experiment to measure the specific heat capacity of a substance	136
CIE 2.2(d)	describe an experiment to measure specific latent heats for steam and for ice	137
CIE 2.3(a)	describe experiments to demonstrate the properties of good and bad conductors of heat	104
CIE 2.3(c)	describe experiments to show the properties of good and bad emitters and good and bad absorbers of infra-red radiation	106
CIE 3.3	measure the speed of sound in air	95
CIE 4.1	describe an experiment to identify the pattern of field lines round a bar magnet	150
CIE 4.2(a)	describe simple experiments to show the production and detection of electrostatic charges	70
CIE 4.2(e)	describe an experiment to determine resistance using a voltmeter and an ammeter	56
CIE 4.5(a)	describe an experiment that shows that a changing magnetic field can induce an e.m.f. in a circuit	164
CIE 4.5(e)	describe an experiment to show that a force acts on a current-carrying conductor in a magnetic field, including the effect of reversing: (i) the current (ii) the direction of the field	164
CIE 4.5(e)	describe an experiment to show the corresponding force on beams of charged particles	159

At the end of this chapter, you will find some practice questions to help you test your knowledge and understanding of practical work and investigations.

Planning investigations

This section covers:

- planning an investigation
- identifying factors to control and measure
- selecting apparatus
- drawing diagrams of apparatus
- describing your method
- safety precautions.

Planning an investigation

In the examinations you may be asked to describe how you would carry out an investigation to find out how one thing depends on another. It helps you to think about an investigation if you can write down a sentence that describes what it is you are investigating. This may be given to you in the question, or you may have to think about this yourself from the information given.

Your sentence or question should mention *two* variables (or factors) only.

- One of these will be the variable you are going to change, sometimes called the **independent variable**.
- The other is the one you are going to measure, sometimes called the **dependent variable** (because its value depends on the independent variable).

Examples

Good examples

How does the current in a wire depend on the voltage?

Here the voltage is the independent variable and the current is the dependent variable.

How does the force on a spring affect its extension?

Here the force is the independent variable and the extension is the dependent variable.

Poor examples

What affects the current in a wire?

What affects the length of a spring?

These examples are not specific enough. They can be turned into good examples by thinking about the factors that *could* affect the current or the spring, and deciding which one to investigate.

Variables and fair tests

Once you have decided on the question you are investigating, you have an independent variable and a dependent variable. You also need to consider what other factors may affect your results. These are the **control variables**.

Control variables need to be kept the same throughout your investigation. This is because if some of these variables change, you will not know which variable has caused any changes in the dependent variable.

A **fair test** is an investigation where only one variable is changed.

> **Example**
>
> If you are investigating the effect of voltage on the current in a wire, you should use the same wire each time in case the length, thickness or material of the wire affects the current. You should also attempt to keep the temperature of the wire the same.

Selecting apparatus

You need to select suitable apparatus to carry out an investigation. Sometimes you will be given a list of apparatus to choose from, or you may have to make your own suggestions.

Remember that you need apparatus to:

- vary the independent variable
- *measure* the independent and dependent variables.

You may also need other apparatus to:

- help to keep the control variables constant
- allow you to carry out the investigation safely.

> **Example**
>
> To investigate how the acceleration of a trolley depends on the force pulling it, you need:
>
> - a trolley and somewhere for it to run
>   - trolley
>   - ramp
> - a means of applying a force to the trolley
>   - string
>   - pulley
>   - masses
> - a way of measuring the acceleration
>   - light gates or ticker tape
> - a way of making sure that masses do not drop on your feet
>   - a box beneath the masses.

You will usually be asked to explain *why* you have chosen particular pieces of apparatus, so don't forget to answer this part of the question as well. Sometimes I will use an ammeter to measure the current is fine, but sometimes you may need more detail.

> **Example**
>
> If you are expecting the current you are measuring to be very small, you would say:
>
> I will use a milliammeter to measure the current, because I expect the current to be very small and a normal ammeter may not give a reading for very small currents.

There is more on measuring instruments on page 197 onwards.

Drawing apparatus

You may be asked to draw diagrams of the apparatus needed for a particular experiment or investigation, to complete a diagram or to label apparatus on a diagram. You don't need to be an artist to do this well! Just make sure your diagram is clear, and that the apparatus is labelled correctly. Remember that you show items such as beakers as a cross-section, as shown in Fig. 8.01.

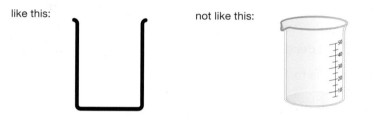

Fig. 8.01: Draw apparatus in cross-section

You may be asked to draw circuit diagrams, so make sure you know all the symbols given in Appendix 2 (see page 242).

Range and number of readings

In most investigations you will be trying to find out if there is a relationship between two variables. Relationships are easiest to see if you plot a graph of your results.

To plot a good graph, you need to use at least five different values of your independent variable. These should have a wide **range** of values. There are likely to be small errors in most of the measurements you make. If there are large differences between the values of the independent variable, the errors in measurements are likely to be a smaller proportion of the values and so you will get more accurate results.

> **Example**
>
> If you are measuring the resistance of different lengths of wire, it would be better to use lengths of 10, 20, 30, 40 and 50 cm, rather than 1, 2, 3, 4 and 5 cm.

You may need to carry out some **preliminary tests** to work out the best values or range of values to use for your independent variable. You do not need to take a lot of readings in preliminary tests – just enough to show that the range of the independent variable you are suggesting will produce a measurable change in the dependent variable.

Physics A* Study Guide

Describing your method

If you are asked to describe a method, take a minute to think about a sensible order in which to describe things. Also remember that in an examination you have a limited amount of time and also a limited amount of space in which to write.

Try to explain your method in a concise manner. If you have already drawn a diagram of the apparatus, or if the question includes a diagram, do not waste time describing what the diagram shows.

Worked example

You are given the following apparatus. Describe how you would use this apparatus to find out how the resistance of a thermistor depends on temperature.

- power supply
- ammeter
- voltmeter
- thermistor

- connecting wires
- beaker
- water
- thermometer

- tripod and gauze
- Bunsen burner

Answer

I would put the thermistor, power supply and ammeter in a series circuit, and connect the voltmeter across the thermistor. I would put the beaker on the tripod and fill it with water. I will put the thermistor into the water, taking care not to get the rest of the apparatus wet. I will switch on the power supply. I will use the thermometer to measure the temperature of the water and read the ammeter and voltmeter. I will light the Bunsen burner and write down the temperature, current and voltage for every 5 °C increase in temperature. I will use the ammeter and voltmeter readings to work out the resistance of the thermistor at each temperature, and plot my results on a graph.

← This part of the answer is describing how to set up the apparatus. You need to describe this here because it is not given in the question.

← This describes how you are going to vary the independent variable (the temperature) and what other measurements you are going to make.

← This describes what you are going to do with your data.

Safety precautions

The safety precautions you need to take for a practical task or investigation depend on the apparatus you are using and what you will be doing with it.

Here are some examples of safety precautions you may need to take when carrying out investigations in physics.

Practical task	Hazard	Precaution
using weights and masses	weights may fall on feet	place a box under the weights – this stops you standing with your feet directly under the weights
stretching wires, springs or elastic bands	the wire or spring may snap and the ends fly about	wear eye protection so that anything flying about cannot harm your eyes
using electricity	electric shock	use a low-voltage power supply keep water away from the circuit
	wires may become hot	allow wires to cool down before touching them
heating	burns from hot objects	handle hot objects with tongs, or allow them to cool down before touching them
using water	slipping on spilt water	mop up any spills straight away

Using measuring instruments

This section covers:

- reading scales
- zero errors
- ways of measuring different quantities.

Reading scales

This section isn't trying to teach you *how* to read a simple scale – you already know that. It is just pointing out a few things to bear in mind.

Fig. 8.02 shows the scale of a forcemeter. Each division represents 0.1 N. However, you can clearly see whether the pointer is on a division, or between two divisions, so you can read the scale to the nearest 0.05 N. In this case, the scale is reading 2.35 N.

Fig. 8.02: The scale on a forcemeter

The reading of 2.35 N on the forcemeter is given to 3 **significant figures**. If you can read a scale to this number of significant figures, then all your data should be recorded to the same number of significant figures. So, if the forcemeter is reading exactly 2 N, you would write this as 2.00 N, to show how precise your data is.

There is more on significant figures on page 201.

Zero errors

A **zero error** occurs if a measuring instrument is not reading zero when it is not measuring anything. If you discovered that your instrument was not set up properly, ideally you would take the measurements again with apparatus that was working correctly. However, this is often not possible, so you should correct all your data for the zero error.

Example

When you have finished taking a series of measurements using a forcemeter, you notice that the forcemeter is reading 2 N when there is no force on it.

You could take the measurements again with a forcemeter that is working correctly, or you could subtract 2 N from all your readings.

Units

Measurements have units! It is no use saying that something is 3 long. You could mean 3 mm, 3 cm, 3 m or even 3 km!

Every time you write down a measurement, include units, or check that the units are given at the top of the column if you are putting data into a table.

Measuring very small things

How do you measure the thickness of a sheet of paper, the mass of one square of graph paper or the swing of a pendulum that lasts for only a couple of seconds?

The answer is to measure many of the items together, and then divide by the number of items measured.

- To measure the thickness of a sheet of paper, measure 10 or 50 or even 100 sheets together and then divide by the number of sheets you have measured.
- To find the mass of one square of graph paper, find the mass of 10 sheets of paper, and then divide by the total number of squares on these sheets.
- To find the time for one swing of a pendulum, time 10 swings and then divide the result by 10.

Measuring lengths

The instrument you need for measuring a length depends on the size of the object being measured.

- For long distances (such as when measuring the speed of sound, or the average running speed over 50 m), use a measuring tape.
- For distances up to a metre, use a metre rule.
- For distances from a few millimetres up to 30 cm, use a ruler.
- For measurements below a few millimetres, use a **micrometer** (also called a **micrometer screw gauge**). A micrometer can measure small distances very accurately. It has a **vernier scale**. This is read by combining information from two scales, as shown in Fig. 8.03.

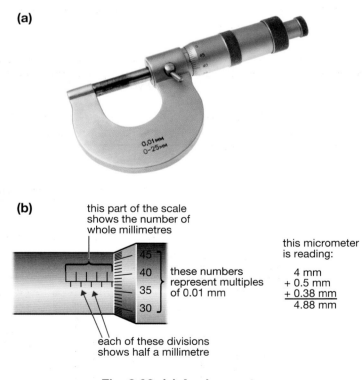

(a)

(b)

this part of the scale shows the number of whole millimetres

these numbers represent multiples of 0.01 mm

this micrometer is reading:

4 mm
+ 0.5 mm
+ 0.38 mm
4.88 mm

each of these divisions shows half a millimetre

Fig. 8.03: (a) A micrometer
(b) How to read a vernier scale

Measuring volumes of liquids

The volume of a liquid can be measured using a **measuring cylinder**.

Water and solutions based on water will form a **meniscus** at the surface. The volume is the reading on the scale at the *bottom* of the meniscus.

To get an accurate reading, your eye should be level with the liquid in the measuring cylinder.

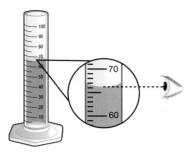

Fig. 8.04: Measuring a volume using a measuring cylinder

Measuring volumes of solids

The volume of a solid object with a regular, rectangular shape can be found by measuring its dimensions and multiplying them together.

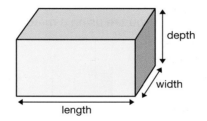

Fig. 8.05: The volume of a rectangular block is found by multiplying its dimensions togeter

The volume of a small, irregular, solid object can be found by dropping it into a measuring cylinder with some water in it. The difference between the readings on the scale before and after the object was added to the water is the volume of the object.

The volume of a larger irregular object can be found using a displacement can. This is filled with water to the level of the bottom of the spout. The object is carefully put into the water and the volume of the displaced water is measured. This is the same as the volume of the object.

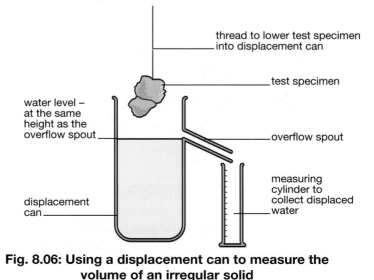

Fig. 8.06: Using a displacement can to measure the volume of an irregular solid

Measuring force and mass

Remember that weight is a force. The weight of an object can be found using a **forcemeter** (also called a **newton meter**). Forcemeters are also used to measure pulling or pushing forces.

A **balance** is used to find the mass of an object. In reality, a balance is measuring the weight of the object, but the scale is calibrated to give you the mass in grams or kilograms.

Measuring time

A **stopclock** or stopwatch can be used for measuring time. Times that you measure by turning the stopclock on and off yourself may have errors due to your reaction time. You will get more reliable results by timing the same event several times and taking a mean.

You may also use a stopclock to help you to record other variables at regular intervals, such as finding out how quickly a beaker of hot water cools down. When you do this, do not stop the clock while taking each reading. The zero measurement is the time at which you start the clock running, and the clock is left running continuously for the whole experiment.

Measuring temperature

Temperature is measured using a **thermometer** or **temperature probe**.

Instruments that measure temperature need to be at the same temperature as the thing they are measuring if they are to provide an accurate measurement. Make sure the reading on the thermometer or probe has stopped changing before recording the temperature.

Some thermometers are filled with mercury. If you are using a mercury thermometer, the reading should be taken from the top of the meniscus.

Fig. 8.07: This mercury thermometer is reading 16 °C

Measuring speed

Speed can be measured:

- by measuring a distance and a time and calculating the speed
- using ticker tape
- using light gates and a datalogger to measure distance and time, or to measure speed directly.

See pages 5 and 9 for details of how to use ticker timers.

Electrical measurements

You should know how to use ammeters and voltmeters from your work on electric circuits.

Some points to remember are:

- Ammeters are always placed in a circuit in series.
- Voltmeters are always placed in parallel to a component.
- If very small currents are being measured, use a milliammeter
- If very small voltages are being measured, use a **galvanometer**.

Obtaining and presenting data

This section covers:

- taking repeated measurements
- recording data
- processing data
- plotting graphs
- finding the gradient and intercept of a graph.

Repeating measurements

All measurements are subject to errors. The way to obtain more reliable data in investigations is to repeat the measurements.

Inspect the measurements for **anomalous data** (results that don't seem to fit the pattern) and try to explain how these anomalies occurred.

Calculate mean results, ignoring any anomalous data, and use the mean results to draw a conclusion.

Line graphs can sometimes be used as an alternative to repeating the same readings. For example, if you are trying to find out the best kind of insulation to keep a beaker of water hot, you might measure the temperature of the water at the start and end of a 10-minute period. To check your data you would have to repeat the same experiment several times with the same insulation, and use a mean of your results. You would then have to do the whole experiment several times more with each different type of insulation.

A much quicker way of doing the experiment is to record the temperature of the water every minute for 10 minutes and then draw a graph of temperature against time. You only need to do this once for each type of insulation. You can then spot any anomalous results because all the temperatures for each type of insulation should lie on a smooth curve.

Recording data

A table is almost always the most appropriate way to record the results of an investigation. Remember:

- the left-hand column is usually your independent variable, or time (for cooling curves or similar investigations)
- each column should have a heading saying what the quantity is (e.g. time, temperature, current)
- the top of each column should also include the units
- you can include columns for 'processed' data – this includes means, but also includes quantities that you need to calculate from other data, such as speed or resistance.

Processing data and significant figures

Processing your data means carrying out calculations. This will usually be calculating a mean, or calculating quantities such as density, speed or resistance, which need to be worked out from other data.

When you are carrying out calculations, think about the number of significant figures in your answers. Your answer should not have more significant figures than the numbers you used to calculate the answer.

Example

You have measured a current of 0.23 A through a resistor, and a voltage of 1.4 V.

Calculating the resistance (resistance = $\dfrac{\text{voltage}}{\text{current}}$) gives a value of 6.086956522 Ω.

If you wrote down this value in your table of results it would imply that your measurement of resistance was accurate to within 10^{-9} Ω, which is clearly not the case! The two measurements you used to calculate resistance each have 2 significant figures, so you should only write down the resistance to 2 significant figures. In this case, rounding would give you a value of 6.1 Ω.

Plotting graphs

An exam question may ask you to plot a graph. Remember the following points when plotting graphs (some may already be done for you in the question).

- Your independent variable (or time, for experiments such as cooling curves) usually goes along the horizontal axis.
- You should choose scales to make use of all the graph paper, but don't use odd scales such as 3 degrees per centimetre – make sure it is still easy to work out where your points should go.
- Each axis should have a label, including units.
- Plot the points with small crosses or dots.
- It is usually most appropriate to draw a line or curve of best fit, ignoring any obviously anomalous results. (Fig. 8.08(a) shows an exception to this rule.)

(a)   (b)

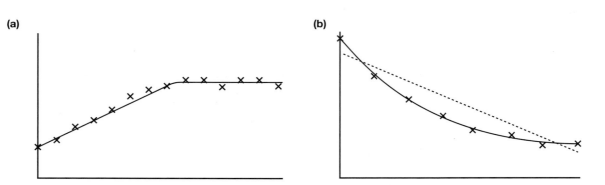

Fig. 8.08: (a) This graph needs two lines of best fit joined with a curve
(b) This graph needs a curve through the points

Interpolation and extrapolation

You may sometimes be asked to take readings from a graph. Fig. 8.09 reminds you how to do this.

- If the reading you are asked for is *between* two plotted points on the graph, you are being asked to **interpolate**.
- If the value is *beyond* any points plotted, you are being asked to **extrapolate**.

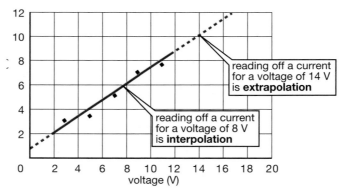

Fig. 8.09: Interpolation and extrapolation

Gradients and intersections

The **gradient** of the line on a graph is a measure of its slope. The **intersection** is the value at which the line crosses one of the axes.

You calculate a gradient using two points on the line. These can be actual data points, but only if *the data points lie on the line of best fit*. Otherwise choose two points anywhere on the line. Make it easy for yourself by choosing points that are easy to read off at least one axis.

$$\text{gradient} = \frac{\text{vertical change}}{\text{horizontal change}}$$

Worked example

Find the gradient of the line on the graph below.

Answer

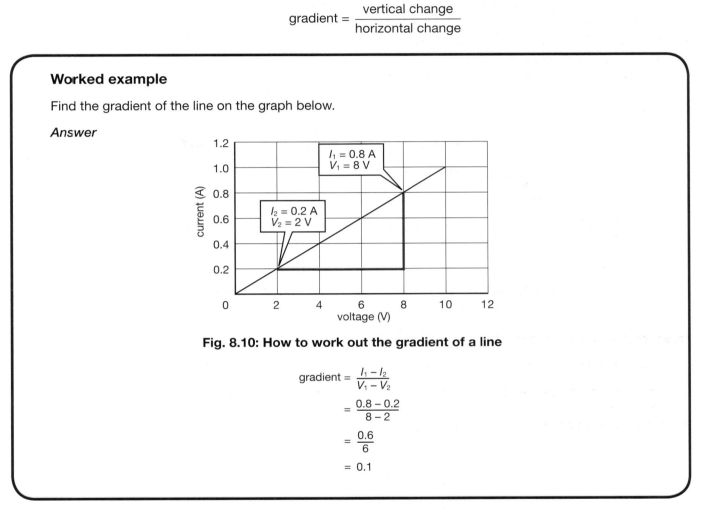

Fig. 8.10: How to work out the gradient of a line

$$\text{gradient} = \frac{I_1 - I_2}{V_1 - V_2}$$

$$= \frac{0.8 - 0.2}{8 - 2}$$

$$= \frac{0.6}{6}$$

$$= 0.1$$

A line that slopes in the opposite direction to the one in Fig. 8.10 has a *negative gradient*. Use exactly the same method as before, remembering to keep the values in the same order on the top and bottom of the equation. This should automatically give you a negative value for the answer.

TIP Check that you have carried out the calculation of gradient correctly by checking the sign.

- A line sloping from bottom left to top right has a positive gradient.

- A line from top left to bottom right (sloping 'downwards') has a negative gradient.

Conclusions and evaluations

This section covers:

- interpreting graphs
- drawing conclusions
- evaluating methods.

Proportionality

Part of being able to write a conclusion to an investigation is to be able to describe in words what a graph of your results shows.

Fig. 8.11 shows two graphs obtained from an experiment to investigate the effect of the load on the length of a spring. On Graph A the length is plotted against the load, and on Graph B the extension is plotted against the load.

- For both graphs you can say that the length (or extension) increases as the load increases.
- You might get *more marks* if you said that there is a **linear relationship** between length (or extension) and load. You are not just saying that the line goes up to the right, but that it is a straight line going up to the right.
- For Graph B, you can also say that the extension is **proportional** to the load.

You can only say that there is a proportional relationship if the straight line on the graph passes through the **origin** (the (0,0) point). This is sometimes referred to as being **directly proportional** to make it clear that you are not talking about an inversely proportional relationship (see below).

A graph like this shows a relationship of the form $y = k \times x$ (or $y = kx$), where k is a constant.

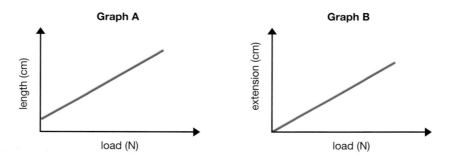

Fig. 8.11: Results of an investigation into springs

Inverse proportionality

If the value of one variable decreases when another increases, one may be **inversely proportional** to the other. A graph showing inverse proportionality is a curve, not a straight line.

Look at Fig. 8.12.

- All you can say about Graph C is that *v* decreases as *w* increases.
- Graph D may show inverse proportionality, but it is difficult to judge whether the curve is exactly the right shape to show this.
- In Graph E, *y* has been plotted against $\frac{1}{x}$. This graph is a straight line that passes through the origin. It shows that *y* is proportional to $\frac{1}{x}$, which is the same as saying that *y* is inversely proportional to *x*.

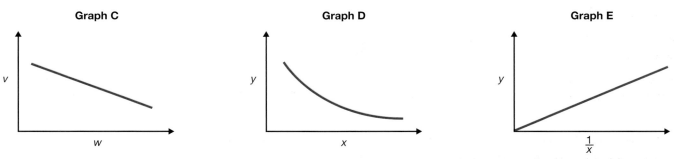

Fig. 8.12: Inverse proportionality is only demonstrated by Graph E

Graph E shows a relationship of the form $y = \frac{k}{x}$, where *k* is a constant.

Drawing a conclusion

The conclusion to an investigation is a statement of what has been found out. Give a justification for your conclusions if you can.

Worked example

This graph shows the results of an investigation to find out how the current through a diode changes with the voltage across it.

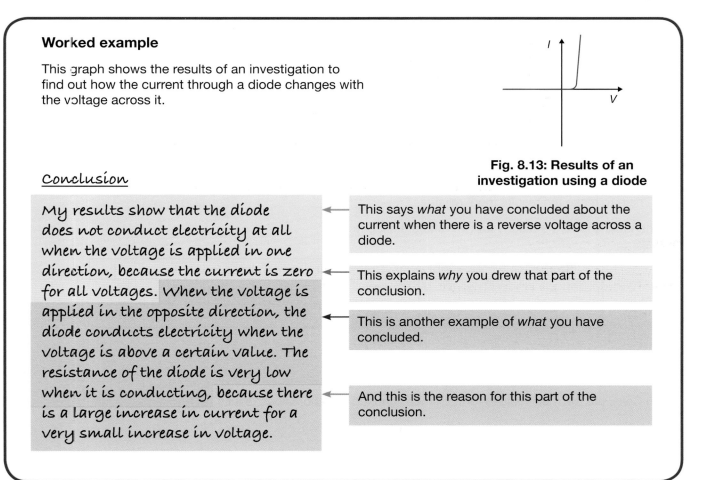

Fig. 8.13: Results of an investigation using a diode

Conclusion

My results show that the diode does not conduct electricity at all when the voltage is applied in one direction, because the current is zero for all voltages. When the voltage is applied in the opposite direction, the diode conducts electricity when the voltage is above a certain value. The resistance of the diode is very low when it is conducting, because there is a large increase in current for a very small increase in voltage.

This says *what* you have concluded about the current when there is a reverse voltage across a diode.

This explains *why* you drew that part of the conclusion.

This is another example of *what* you have concluded.

And this is the reason for this part of the conclusion.

Evaluating

You may be asked to evaluate a method or a set of results.

If you are asked to evaluate a method, consider the following:

- Does the method described include repeated measurements? If not, suggest that more measurements are taken.
- Could a more accurate or precise measuring instrument have been used? If so, suggest a better instrument or way of measuring.
- Could some of the experimental conditions have been controlled better? If so, suggest how this could be done.

If you are asked to evaluate a set of results, consider the following:

- Are the repeated results all similar to one another? If so, the evidence is probably reliable.
- Do the plotted points all lie very close to the line or curve of best fit on the graph? If so, the evidence is probably reliable.

Practice questions

1. A student is using the apparatus shown below to investigate the speed at which different shapes fall through oil.

(a) The student did some preliminary tests to find the best liquid to use. Suggest why she chose oil rather than water. **(3)**

(b) Suggest how the lines on the measuring cylinder are used during the investigation, and how they could help to make the results more reliable. **(2)**

The table shows the results for two different shapes.

Shape	Time to fall between lines (s)				
A	3.12	2.87	3.05	6.15	2.91
B	3.54	3.89	3.96	4.11	3.47

(c) (i) Which reading is an anomalous result? **(1)**

    (ii) Suggest how this error was made. **(1)**

(d) Calculate the mean values for the times, ignoring the anomalous results. **(5)**

2. A student sets up the following apparatus and uses it to investigate the resistance of a set of wires.

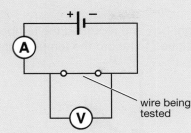

wire being tested

The student gets the following results:

Voltage (V)	Current (A)
0	0
1	0.07
2	0.12
3	0.19
4	0.28
5	0.31
6	0.39

(a) Plot a graph to show these results, with voltage on the horizontal axis. **(4)**

(b) What is the current through the wire when the voltage is 1.5 V? **(1)**

(c) Use your answer to part (b) to work out the resistance of the wire. **(2)**

(d) Explain why working the resistance out from the graph like this provides a more reliable result than using one of the individual results. **(2)**

The student uses the same method to find the resistances of wires of different cross sections. The graph shows his results.

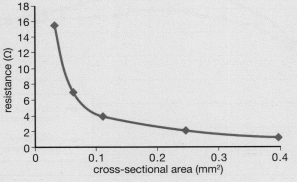

(e) Suggest one variable he should keep the same to make his test fair. **(1)**

He writes this conclusion.

The resistance of a wire decreases as its cross-sectional area increases. The resistance is inversely proportional to the cross-sectional area.

(f) (i) Explain why this conclusion cannot be justified based only on the graph shown. **(1)**

   (ii) Suggest a different graph that the student could plot to check that his conclusion was correct. **(2)**

3. Two groups of students are investigating how the colour of a surface affects the rate at which it absorbs and emits infrared radiation. The bulb of one thermometer is covered in white paint, and the bulb of the other is covered in black paint.

They switched on the bulb in the light box and recorded the temperature every minute for 10 minutes. Then they switched the bulb off and continued to record the temperature every minute for a further 10 minutes.

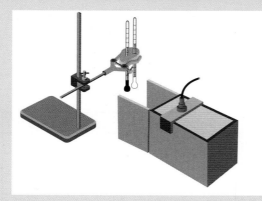

Group A are planning to use the apparatus shown. One member of the group will monitor a stopclock and call out every minute. Two other members of the group will read one thermometer each, every minute.

Group B think that it would be better to have the same person reading both thermometers, as people might read the scale on the thermometer differently. They will carry out the experiment with the white-painted thermometer first, and then repeat it with the black-painted one.

(a) Describe three variables that should be kept the same to ensure the investigation is a fair test. **(3)**

(b) Explain why Group A's method will produce more reliable results. **(2)**

The graph shows the results of Group A's experiment.

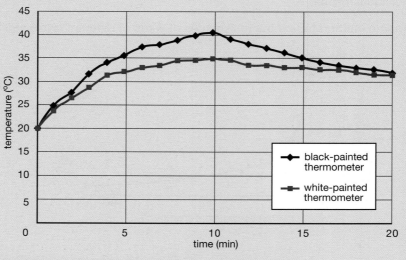

(c) What mistake have this group made in drawing their graph? **(1)**

(d) What conclusion can you draw from this graph about:

    (i) the rate at which different colours absorb infrared radiation **(2)**

    (ii) the rate at which different colours emit infrared radiation? **(2)**

(e) Group A do not need to repeat their experiment to obtain reliable results. Explain why. **(1)**

(f) Explain how these results can provide a reliable conclusion about the rate at which different surfaces emit radiation, even though the thermometers did not both reach the same maximum temperature. **(1)**

Answers

1 Forces and motion

A Movement and position

Review questions (page 11)

1. distance = 2 m, time = 1.5 s

$$\text{speed} = \frac{\text{distance}}{\text{time}} = \frac{2 \text{ m}}{1.5 \text{ s}} = 1.33 \text{ m/s}$$

2. $\text{time} = \frac{\text{distance}}{\text{speed}} = \frac{200 \text{ m}}{6 \text{ m/s}} = 33.3 \text{ s}$

3. (a) $\text{acceleration} = \frac{\text{change in velocity}}{\text{time}} = \frac{(7 \text{ m/s} - 2 \text{ m/s})}{2 \text{ s}} = \frac{5}{2} = 2.5 \text{ m/s}^2$

   (b) change in velocity = acceleration × time = 2 m/s² × 3 s = 6 m/s

   initial speed = 0.5 m/s, final speed = 0.5 m/s + 6 m/s = 6.5 m/s

4. (a) It is stationary for 5 seconds, then it accelerates to 8 m/s over the next 10 seconds and then travels at a constant 8 m/s for 10 seconds.

   (b) $\text{acceleration} = \frac{\text{change in velocity}}{\text{time}} = \frac{(6 \text{ m/s} - 8 \text{ m/s})}{20 \text{ s}} = \frac{-2 \text{ m/s}}{20 \text{ s}} = -0.10 \text{ m/s}$

   (Remember that a deceleration is a negative acceleration.)

   (c) distance = area under graph = ½ × 10 s × 8 m/s = 40 m

Practice questions (page 12)

1. (a) Between 2.5 and 3 hours after starting **(1)** because the line is horizontal/she is not moving/the distance is not changing **(1)**

   (b) Between 1.5 and 2.5 hours after starting **(1)** because she is moving slower here/the gradient of the graph is shallower here **(1)**

   (c) $\text{speed} = \frac{\text{distance}}{\text{time}}$ **(1)**

   $= \frac{7.5 \text{ km}}{1.5 \text{ hours}}$ **(1)**

   = 5 km/h   **(1 for correct number, 1 for correct units)**

   (You would also be correct if you had read values from intermediate points on the graph, such as taking a distance of 5 km at a time of 1 hour.)

2. (a) Its velocity is increasing. **(1)**

   (b) $\text{acceleration} = \frac{\text{change in velocity}}{\text{time}}$ (or $a = \frac{(v - u)}{t}$) **(1)**

   $= \frac{(0 \text{ m/s} - 10 \text{ m/s})}{60 \text{ s}}$ **(1 for substituting the correct numbers)**

   $= -0.167 \text{ m/s}^2$ **(1 for correct answer, 1 for correct units)**

   (c) Work out the area under the graph (or the area between the graph and the time/horizontal axis). **(1)**

(d) For 0 to 30 seconds, area $= \frac{1}{2} \times 30 \text{ s} \times 5 \text{ m/s}$ **(1)**

$$= 75 \text{ m} \textbf{ (1)}$$

For 30 to 60 seconds, area $= 30 \text{ s} \times 5 \text{ m/s}$ **(1)**

$$= 150 \text{ m} \textbf{ (1)}$$

Total distance travelled $= 225 \text{ m}$ **(1)**

B Forces, movement, shape and momentum

Review questions (page 32)

1. Any two from: air resistance, water resistance, two surfaces rubbing together.

2. (a) 5 N to the right

   (b) 17 N to the left (If you have put 5 N and –17 N, or –5 N and 17 N, that is fine. The important thing is that your answer makes it clear that the two resultant forces are in opposite directions.)

3. Your scale drawing should look something like this:

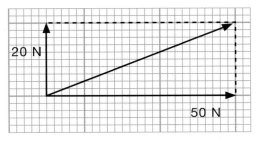

   The resultant is 54 N at 22° from the 50 N force.

4. (a) $a = \dfrac{F}{m} = \dfrac{10 \text{ N}}{2 \text{ kg}} = 5 \text{ m/s}^2$

   (b) $m = \dfrac{F}{a} = \dfrac{15 \text{ N}}{5 \text{ m/s}^2} = 3 \text{ kg}$

5. weight = mass × g = 10 kg × 9.81 N/kg = 98.1 N

6. (a) zero

   (b) It is greatest when they first leave the aeroplane, and then decreases until it is zero when they are at terminal velocity.

   (c) Their weight is constant, but air resistance increases as they get faster. The air resistance acts in the opposite direction to the weight, so as they get faster the resultant downward force gets smaller.

7. momentum = mass × velocity = 20 kg × 50 m/s = 1000 kg m/s

8. velocity $= \dfrac{\text{momentum}}{\text{mass}} = \dfrac{75 \text{ kg m/s}}{15 \text{ kg}} = 5 \text{ m/s}$

9. It will be in the same direction as the original movement. It will be less than the velocity of the single trolley, because the overall mass is bigger.

10. force $= \dfrac{\text{change in momentum}}{\text{time taken}} = \dfrac{20 \text{ kg m/s}}{4 \text{ s}} = 5 \text{ N}$

11. Any two from: crumple zone, air bags, seat belt.

12. You can produce a greater moment with a longer spanner.

13. moment of boy = force × distance = 400 N × 2 m = 800 Nm

Moment of girl must be the same magnitude but in the opposite direction to balance the boy.

$$\text{distance} = \frac{\text{moment}}{\text{force}} = \frac{800 \text{ N m}}{500 \text{ N}} = 1.6 \text{ m}$$

She must sit 1.6 m from the pivot.

14. When you start, almost all of your weight is on the bank on the side where you start. As you walk across the bridge, the force on this bank gets smaller and the force on the far bank gets greater, until all the force is on the far bank as you finish crossing the stream.

15. It has a heavy base, which makes its centre of mass low. It has a wide base, so it has to be tilted a long way before it will fall over.

Practice questions (page 33)

1. (a) It takes time for a car to stop **(1)** and it continues to move along the road while it is stopping. **(1)** The faster the car is going, the more distance it covers while it is stopping. **(1)**

(b) thinking distance and braking distance **(1)**

(c) If it is raining it might be harder for drivers to identify that there is danger ahead, so the thinking distance could increase. **(1)** Braking will be less effective on a wet road, so the braking distance could also increase. **(1)**

2. (a) 200 N **(1)** downwards **(1)**

(b) $\text{acceleration} = \dfrac{\text{force}}{\text{mass}}$ (or $a = \dfrac{F}{m}$) **(1)**

$$= \frac{200 \text{ N}}{70 \text{ kg}} \text{ (1)}$$

$$= 2.86 \text{ m/s}^2 \text{ (1 for correct answer, 1 for correct unit)}$$

(c) 700 N **(1)** The forces on her are balanced (or the upward and downward forces are equal) when she has reached terminal velocity. **(1)**

(d) It would be faster/greater **(1)** because she would have to be going faster for the air resistance to balance her greater weight. **(1)**

(e) (i) $\text{acceleration} = \dfrac{\text{change in velocity}}{\text{time}}$

$$= \frac{(5 - 55)}{3} \text{ (1)}$$

$$= -16.67 \text{ m/s}^2 \text{ (1 for correct number, 1 for minus sign)}$$

(ii) force = mass × acceleration
$$= 70 \text{ kg} \times -16.67 \text{ m/s}^2 \text{ (1)}$$
$$= -1167 \text{ N (1)}$$

3. (a) momentum = mass × velocity
$$= 1500 \text{ kg} \times 30 \text{ m/s (1)}$$
$$= 45\,000 \text{ kg m/s (1 for correct answer, 1 for unit)}$$

(b) total mass of cars after collision = 2500 kg

$$\text{velocity} = \frac{\text{momentum}}{\text{mass}} \textbf{(1)}$$

$$= \frac{45\,000 \text{ kg m/s}}{2500 \text{ kg}} \textbf{(1)}$$

$$= 18 \text{ m/s } \textbf{(1)}$$

(c) No energy is added to or removed from the system. **(1)**

4. momentum before the collision = 0.15 kg × 6 m/s = 0.9 kg m/s **(1)**
momentum after the collision = –0.9 kg m/s
(because the ball is moving at the same speed but in the opposite direction)
change in momentum = 1.8 kg m/s **(1)**

$$\text{force} = \frac{\text{change in momentum}}{\text{time taken}}$$

$$= \frac{1.8 \text{ kg m/s}}{0.005 \text{ s}} \textbf{(1)}$$

$$= 360 \text{ N } \textbf{(1)}$$

5. (a) weight = mass × g
= 1.2 kg × 10 N/kg **(1)**
= 12 N **(1)**

(b) The card on the top of the trolley breaks the beam in the light gate. **(1)**
The datalogger calculates the speed from the length of the card and the time that the card took to pass through the light gate. **(1)**

(c) Find the speed change by subtracting the first speed from the second one. **(1)**
The datalogger also records the time at which the two speed measurements were taken. **(1)**
Calculate the acceleration from change in speed ÷ time. **(1)**
(You may not have written your answer in exactly the same way, but give yourself the marks if your explanation includes the key points above.)

(d) Friction will slow the trolley down. **(1)**
Gravity pulling the trolley down the ramp compensates for the effects of friction. **(1)**

(e) The masses on the end of the string are also accelerating. **(1)**
By moving the masses from the trolley to the end of the string, the total mass of the things accelerating is kept the same. **(1)**

(Well done if you got part (e) correct – this was a hard question!)

6. (a) B **(1)** because this is the distance perpendicular (or at right angles) to the force. **(1)**

(b) This increases their moment about the pivot **(1)** and reduces the force needed on rope A to open the bridge. **(1)**

(c) The moment depends on the perpendicular distance between the force and the pivot **(1)** so a given force produces a bigger moment if the rope is attached at X. **(1)**

7. (a) The wider base makes it more stable **(1)** because it would have to be tilted further before its centre of gravity was outside the base. **(1)**

(b) It will make it less stable **(1)** because it will raise its centre of gravity **(1)** so the glass does not have to be tilted so far before the centre of gravity is no longer over the base. **(1)**

8. (a) The force needed to extend a spring is proportional to its extension. **(1)**

(b) The spring behaves elastically if it is not stretched beyond the limit of proportionality **(1)** so it returns to its original length when the force is removed. **(1)**
This is important as you must be able to stretch and pull the bolt back many times. **(1)**

C Astronomy

Review questions (page 40)

1. (a) Sun   (b) planet   (c) Sun

2. First convert the time into seconds:

   $T = 27.3 \times 24 \times 60 \times 60 = 2\ 358\ 720$ s

   Change the distance into metres:

   $r = 384\ 399\ 000$ m

   $$v = \frac{2 \times \pi \times r}{T}$$

   $$= \frac{2 \times 3.14 \times 384\ 399\ 000 \text{ m}}{2\ 358\ 720 \text{ s}}$$

   $$= 1023 \text{ m/s}$$

   (You may get 1024 if you have used the π button on your calculator instead of entering 3.14. Either would be marked correct in an exam.)

3. $T = \dfrac{2 \times \pi \times r}{v} = \dfrac{2 \times 3.14 \times 5.79 \times 10^{10} \text{ m}}{48\ 000 \text{ m/s}} = 7.57 \times 10^{6}$ s

   (This is about 88 days, just for interest.)

Practice questions (page 40)

1. (a) gravitational field strength $= \dfrac{\text{weight}}{\text{mass}}$

   $$= \frac{431 \text{ N}}{319 \text{ kg}} \textbf{ (1)}$$

   $$= 1.35 \text{ N/kg } \textbf{(1)}$$

   (b) The Moon has a bigger mass than Titan **(1)** because its gravitational field strength is greater. **(1)**

   (c) radius $= 355\ 000\ 000$ m **(1)**

   period $= 5.9 \times 24 \times 60 \times 60$ s $= 509\ 760$ s **(1)**

   orbital speed $= \dfrac{2 \times \pi \times r}{T}$

   $$= \frac{2 \times 3.14 \times 355\ 000\ 000}{509\ 760 \text{ s}} \textbf{ (1)}$$

   $$= 4373 \text{ m/s } \textbf{(1)}$$

2 Electricity

A Mains electricity

Review questions (page 49)

1. current $= \dfrac{\text{power}}{\text{voltage}} = \dfrac{1200 \text{ W}}{110 \text{ V}} = 10.9$ A

2. $t = \dfrac{E}{(I \times V)} = \dfrac{90 \text{ J}}{(0.5 \text{ A} \times 12 \text{ V})} = 15$ s

3. power = 1000 W

$$I = \frac{P}{V} = \frac{1000 \text{ W}}{230 \text{ V}} = 4.35 \text{ A}$$

Use a 5 A fuse.

Practice questions (page 49)

1. (a) current = $\dfrac{\text{power}}{\text{voltage}}$

$$= \frac{11 \text{ W}}{230 \text{ V}} \text{ (1)}$$

$$= 0.05 \text{ A (1)}$$

(b) energy = current × voltage × time (or power × time) **(1)**

= 0.05 A × 230 V × 60 × 60 **(1)**

(remember that there are 60 × 60 seconds in an hour)

= 41 400 J (or 39 600 J if you have multiplied 11 W by 3600 s – the difference is because the 0.05 A calculated in part (a) is rounded to 1 s.f.) **(1)**

(c) 3 A **(1)**

This is the lowest rating above the current used by the lamp. **(1)**

(d) If there is a fault and the outer casing becomes live, a user could get a shock. **(1)**

The earth wire provides a low-resistance path from the casing to earth **(1)** and so a large current flows if a fault occurs. **(1)**

This melts the fuse and stops electricity flowing to the appliance. **(1)**

(e) It will be safe if it is double insulated **(1)** which means that there are no metal parts that the user can touch. **(1)**

B Energy and potential difference in circuits

Review questions (page 66)

1. (a) The current will increase.

(b) The current will increase.

2. (a) 20 Ω

(b) $\dfrac{1}{R} = \dfrac{1}{R_1} + \dfrac{1}{R_2} = \dfrac{1}{10} + \dfrac{1}{10} = \dfrac{2}{10} = 0.2$

$R = \dfrac{1}{0.2} = 5 \text{ Ω}$

3. 6 V (if the bulbs are the same, each one gets half the voltage)

4. A component that converts changes in a physical quantity to an electrical signal.

5. $R = \dfrac{V}{I} = \dfrac{12 \text{ V}}{2 \text{ A}} = 6 \text{ Ω}$

6. $I = \dfrac{Q}{t} = \dfrac{100 \text{ C}}{0.5 \text{ s}} = 200 \text{ A}$

7. (a) thermistor and LDR

(b) AND

Practice questions (page 66)

1. (a) So they can be switched on and off independently. **(1)**

   (b) So if one breaks the other will still work. **(1)**

   (c) Both lights would be dimmer **(1)** because the current would be lower (or the resistance would be higher). **(1)**

2. (a) $\dfrac{1}{R} = \dfrac{1}{2500} + \dfrac{1}{2500}$ **(1)**

   $= 0.0008$

   $R = 1250\ \Omega$ **(1)**

   (b) $I = \dfrac{V}{R}$ **(1)**

   $= \dfrac{230\ V}{2500\ \Omega}$ **(1)**

   $= 0.09\ A$ **(1)**

3. (a) The graph passes through the origin **(1)** as no current can flow when there is no voltage across the filament. **(1)**

   The current is negative when the voltage is negative **(1)** as the direction in which the current flows depends on the direction in which the voltage is applied. **(1)**

   The gradient of the line gets less as the voltage increases **(1)** as the resistance of the filament increases at higher voltages. **(1)**

   (b) The energy transferred by a current depends on the current and the voltage, as shown by the formula $P = I \times V$. **(1)**
   This energy will be transferred as heat **(1)** so the higher the voltage and current, the higher the temperature of the wire. **(1)** Therefore as resistance increases at higher voltages and currents, the resistance is increasing at higher temperatures. **(1)**

4. (a) Any two of the following points:
   The LED comes on when the increasing resistance across the capacitor has increased the voltage across it to the value at which the LED starts to conduct. **(1)**
   If the resistance of X is less, a higher current will flow initially and the capacitor will charge up and reach this resistance faster. **(1)**
   If the resistance of X is less, X will take a smaller proportion of the 5 V, and so the voltage across BC will reach the required voltage for the LED to conduct sooner. **(1)**

   (b) The capacitor will be charged up **(1)** and when the circuit is switched off the capacitor will discharge. **(1)**
   The potential difference between the positively charged plate on the capacitor and the 0 V line will be above the value needed to light the LED for a few seconds. **(1)**

5. (a) $R = \dfrac{V}{I}$

   $= \dfrac{12\ V}{4\ A}$ **(1)**

   $= 3\ \Omega$ **(1)**

   (b) $Q = I \times t$
   $= 4\ A \times 10\ s$ **(1)**
   $= 40\ C$ **(1 for the correct answer, 1 for the correct unit)**

   (c) 1 volt represents 1 joule transferred per coulomb of charge **(1)**
   so 12 V transfers 12 V × 40 C of energy, which is 480 J. **(1)**

(d) $E = I \times V \times t$
      $= 4\,A \times 12\,V \times 10\,s$ **(1)**
      $= 480\,J$ **(1)**

6. (a) Your sketch should look something like this:

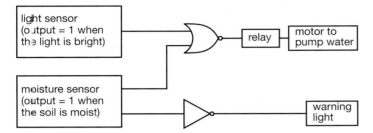

(1 for showing all the sensors and
motor/light on the diagram
1 for a NOR gate
1 for the NOT gate
1 for connecting them up as shown)

(b) The system needs to pump water when the soil is dry and the light is not bright,
so it needs to switch on the relay when both inputs are 0. **(1)**
This can be achieved using a NOR gate. **(1)**
The light needs to be on when the soil is dry, so when the moisture sensor output is 0 the light needs to
be on. **(1)**
A NOT gate can be used to do this. **(1)**

C Electric charge

Review questions (page 72)

1. Your sketch should look like the first part of Fig. 2c.02 (on page 69), but with the arrows pointing inwards.

2. The comb has an electrostatic charge. This induces an opposite charge on the side of the water stream
closest to it, and the water moves towards the comb.

3. So that any electrostatic charge can be evened out or made to flow to earth, which means there will be no
sparks which might cause an explosion.

4. So that the toner powder will only stick to the plate in the correct pattern.

Practice questions (page 72)

1. (a) If the student touches a metal object the static charge on him/her will be conducted away. **(1)**

   (b) The student, the container and the puffed rice all get the same static charge from the
Van de Graaff generator. **(1)**
Like charges repel **(1)** so when enough charge has built up, the pieces of puffed rice are repelled
strongly enough to fly out of the container. **(1)**

   (c) The static charge on the generator induces an opposite charge on the sides of the bubbles
closest to it. **(1)**
Opposite charges attract **(1)** so the bubbles are attracted towards the generator. **(1)**

2. (a) A spark may jump between objects with different static charges **(1)** which could ignite any fuel vapour in
the air. **(1)**

   (b) If the ground can conduct some electricity, a current can flow between objects with different charges **(1)**
which will allow the charges to equalise. **(1)**
If there are no objects with different charges, no sparks will occur. **(1)**

   (c) A vehicle may build up a static charge while it is being used, and the rubber tyres may prevent this being
discharged at the petrol station. **(1)**
So a metal container in contact with the vehicle may have a different charge to the nozzle
(or pump) and a spark may occur. **(1)**
If the container is put on the ground, any difference in charge will be equalised and no spark
will occur. **(1)**
Keeping the metal nozzle in contact with the container will ensure that they both have the
same static charge and so no spark will occur. **(1)**

3. All the drops of insecticide will have a charge of the same sign **(1)** so they will all repel each other. **(1)** This will help the spray to spread out and cover a bigger area (or cover the area more evenly). **(1)**

3 Waves

A Properties of waves

Review questions (page 78)

1. $\lambda = \dfrac{v}{f} = \dfrac{330 \text{ m/s}}{1000 \text{ Hz}} = 0.33 \text{ m}$

2. time period $= \dfrac{1}{\text{frequency}} = \dfrac{1}{50 \text{ Hz}} = 0.02 \text{ s}$

3. The wavelength of sound waves is similar to the width of the door, so the waves will be diffracted by the door way and will spread out beyond it. This does not happen with the light waves, because the wavelength of light waves is very much smaller than the width of the door.

Practice questions (page 79)

1. (a) transverse **(1)**

   (b) speed $= \dfrac{\text{distance}}{\text{time}}$ **(1)**

   $= \dfrac{3\ 000\ 000 \text{ m}}{15\ 000 \text{ s}}$

   $= 200 \text{ m/s}$ **(1)**

   (c) period $= 20$ minutes **(1)**
   $= 20 \times 60$ seconds
   $= 1200$ seconds **(1)**

   (d) $f = \dfrac{1}{T}$ **(1)**

   $= \dfrac{1}{1200 \text{ s}}$

   $= 0.0008 \text{ Hz (or } 8 \times 10^{-4} \text{ Hz)}$ **(1)**

   (e) wavelength $= \dfrac{\text{speed}}{\text{frequency}}$ **(1)**

   $= \dfrac{200 \text{ m/s}}{0.0008 \text{ Hz}}$

   $= 250\ 000 \text{ m}$ **(1)**

2. (a) A, E and F are good places to anchor in a swell **(1)** as the wavelength of the swell is much larger than the harbour entrance **(1)** so there will be very little diffraction. **(1)**

   (b) If the sea is choppy, A and E are the best places to anchor **(1)** as the wavelength of the waves is similar to the width of the harbour entrance **(1)** so diffraction will enable the choppy waves to reach B and F. **(1)**

   (Give yourself the first mark if you included B as well.)

B The electromagnetic spectrum

Review questions (page 82)

1. (a) infrared

    (b) microwaves

    (c) infrared

    (d) visible light (although photos can also be taken using other waves, such as X-ray photographs)

    (e) ultraviolet

    (f) gamma rays

    (g) X-rays

    (h) infrared (in grills, toasters etc.) and microwaves

    (i) visible light

2. Because they are warmer than their surroundings.

3. (a) ultraviolet

    (b) gamma rays, ultraviolet

    (c) microwaves, infrared

Practice question (page 83)

1. (a) red, orange, yellow, green, blue, indigo, violet **(1)**

    (b) infrared **(1)**

    (c) (i) ultraviolet **(1)**

    (ii) skin damage/skin cancer **(1)**
    blindness **(1)**

    (d) They travel at the same speed in a vacuum. **(1)**

    (e) (i) gamma rays **(1)**

    (ii) radio waves **(1)**

    (f) (i) Any two from: making images of the inside of the body, luggage scans at airports, checking the internal structure of metal objects. **(2)**

    (ii) Any two from: sterilising surgical instruments, sterilising/irradiating food, detecting and treating cancer. **(2)**

C Light and sound

Review questions (page 96)

1. 30°

2. $\sin r = \dfrac{\sin i}{n} = \dfrac{\sin 30°}{1.33} = 0.376$

    $r = 22.1°$

3. $\sin c = \dfrac{1}{n}$

$= \dfrac{1}{2.4}$

$= 0.417$

$= 24.6°$ (You may get 24.8° if you round sin c before finding the angle.)

4. (a) The pitch will become higher.

   (b) The loudness will not change.

5. A ray that passes straight through the centre of the lens without changing direction; a ray from the object parallel to the axis of the lens, which is bent to pass through the principal focus on the far side of the lens; a ray from the object that passes through the principal focus on the same side of the lens, which emerges parallel to the lens axis.

6. real, inverted (you cannot say whether the object is magnified or diminished unless you know the actual distance of the object from the lens)

7. (a) less than one focal length from the lens

   (b) virtual, upright, magnified

Practice questions (page 97)

1. (a) Your diagram should look like this:

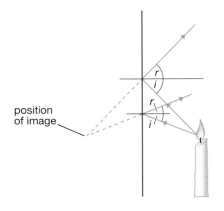

(1 mark for drawing straight lines with arrows to show the direction in which the light is travelling
1 mark for making the angles of incidence and reflection equal
1 mark for showing the position of the image)

   (b) *i* and *r* marked in one of the positions shown. **(1)**

   (c) The angle of incidence is equal to the angle of reflection. **(1)**

2. (a) Your ray diagram should look like this. If you have drawn it accurately, the image will be 7.5 cm from the lens, and will be 1.5 times the size of your object. On this diagram, five small squares on the graph paper represent 1 cm.

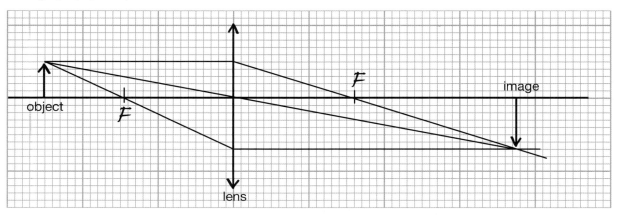

(1 mark for each of the three lines drawn correctly)

(b) magnified **(1)** inverted **(1)** real **(1)**

3. (a) $\sin r = \dfrac{\sin i}{n}$ **(1)**

$= \dfrac{\sin 20°}{0.66}$

$= 0.518$ **(1)**

$r = 31.2°$ **(1)**

(b) $\sin c = \dfrac{1}{n}$ **(1)**

$= \dfrac{1}{1.5}$

$= 0.66$ **(1)**

$c = 41.3°$ **(1)**

(c) Total internal reflection will occur **(1)** because the angle of incidence is greater than the critical angle. **(1)**

(d) The light will pass out of the glass/total internal reflection will not occur **(1)** because the angle of incidence is smaller than the critical angle for light going from glass to water. **(1)**

(e) If the sensor detects infrared light from the emitter that has been reflected by the inner surface of the glass, the windscreen is dry and the wipers are not needed. **(1)**
If it does not detect this light, then the windscreen is wet and the wipers are switched on. **(1)**

4. (a) Analogue signals vary continuously **(1)** whereas digital signals are a series of pulses (or 0s and 1s). **(1)**

(b) When analogue signals are amplified any noise on them is also amplified. This does not happen with digital signals. **(1)**
Digital signals can carry more information, so more TV channels can be broadcast on the same range of frequencies. **(1)**

5. (a) longitudinal **(1)**

(b) amplitude **(1)**
A sketch similar to this: **(1)**

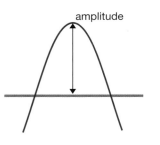

amplitude

(c) period **(1)**
   The *x* axis/horizontal axis on the oscilloscope represents time so the distance between two peaks shows the time between them. **(1)**

(d) 0.282 m **(1)**

(e) To allow for/help her to spot any mistakes. **(1)**

(f) speed = frequency × wavelength **(1)**
   = 1200 Hz × 0.282 m **(1)**
   = 338 m/s **(1)**

4 Energy resources and energy transfer

A Energy transfer

Review questions (page 108)

1. efficiency $= \dfrac{\text{useful energy transferred}}{\text{total energy transferred}} = \dfrac{30\ \text{J}}{50\ \text{J}} = 0.6$ (or 60%)

2. Your Sankey diagram should look like this:

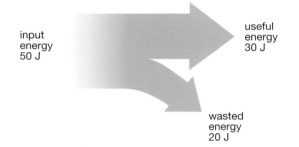

input
energy
50 J

useful
energy
30 J

wasted
energy
20 J

   Remember that the widths of the arrows must be proportional to the amounts of energy they represent.

3. (a) radiation

   (b) convection

   (c) conduction

   (d) radiation

Practice questions (page 109)

1. (a) Your diagram should look something like this:
   chemical energy (in fuel) → thermal/heat energy (in water and steam) → kinetic energy (in moving parts of engine and pulley) → gravitational potential energy (in lifted weight)
   (1 mark for each type of energy to max. 4)

   (b) efficiency $= \dfrac{\text{useful energy output}}{\text{total energy input}}$ **(1)**

   $= \dfrac{5\ \text{J}}{7\ \text{J}}$

   = 0.71 (or 71%) **(1)**

(c) Your Sankey diagram should look like this:

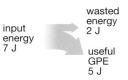

input
energy
7 J

wasted
energy
2 J

useful
GPE
5 J

**(1 mark for drawing the arrows to scale
1 mark for labelling them correctly)**

2. (a) Black is the best colour for absorbing radiation. **(1)**

   (b) (i) The pot is solid, and heat energy is transferred through solids by conduction. **(1)**

   (ii) Metals are good conductors of heat energy. **(1)**

   (c) convection **(1)**
   Convection transfers heat energy when hot fluids rise and set up convection currents. **(1)**
   The lid is designed to stop warm air rising and carrying heat energy away from the box. **(1)**

   (d) (i) conduction **(1)**

   (ii) Any material that contains pockets of air **(1)** as air is a good insulator/poor conductor. **(1)**
   (1 mark if you just said wood/plastic/wool)

B Work and power

Review questions (page 115)

1. $\text{force} = \dfrac{\text{work}}{\text{distance}} = \dfrac{500 \text{ J}}{2.5 \text{ m}} = 200 \text{ N}$

2. $\text{GPE} = m \times g \times h = 5 \text{ kg} \times 9.81 \text{ N/kg} \times 10 \text{ m} = 490.5 \text{ J}$

3. $h = \dfrac{\text{GPE}}{(m \times g)} = \dfrac{300 \text{ J}}{(10 \text{ kg} \times 9.81 \text{ N/kg})} = 3.06 \text{ m}$

4. $\text{KE} = \frac{1}{2} \times m \times v^2 = \frac{1}{2} \times 2 \text{ kg} \times (5 \text{ m/s})^2 = 25 \text{ J}$

5. $v = \sqrt{\dfrac{2 \times \text{KE}}{m}} = \sqrt{\dfrac{2 \times 100 \text{ J}}{0.5 \text{ kg}}} = \sqrt{400} = 20 \text{ m/s}$

6. $\text{power} = \dfrac{\text{work}}{\text{time}} = \dfrac{300 \text{ J}}{2 \text{ s}} = 150 \text{ W}$

Practice questions (page 115)

1. (a) They have both done the same amount of work. **(1)**
   They are both doing work against gravity **(1)** so the distance moved in the direction of the force is the same, as they are both climbing the same distance from the ground to the footbridge. **(1)**

   (b) force = 90 kg × 10 N/kg **(1)**
   = 900 N **(1)**

   work = force × distance **(1)**
   = 900 N × 6 m
   = 5400 J **(1 for the correct number, 1 for the correct unit)**

   (c) 5400 J **(1)**
   The work done in lifting himself and the suitcase to the top of the bridge is the same as the gravitational potential energy gained. **(1)**

(d) power = $\dfrac{\text{work}}{\text{time}}$ **(1)**

$= \dfrac{5400 \text{ J}}{30 \text{ s}}$

= 180 W **(1 for the correct number, 1 for the correct unit)**

2. (a) GPE = $m \times g \times h$ **(1)**

= 0.05 kg × 10 N/kg × 1.5 m **(1)**

= 0.75 J **(1)**

(b) 0.75 J **(1)**

(c) KE = $\frac{1}{2} \times m \times v^2$

$v = \sqrt{\dfrac{2 \times \text{KE}}{m}}$ **(1)**

$= \sqrt{\dfrac{2 \times 0.75 \text{ J}}{0.05 \text{ kg}}}$ **(1)**

$= \sqrt{30}$

v = 5.48 m/s **(1)**

(d) It has gained some gravitational potential energy by going up to the top of the loop **(1)** so its kinetic energy must be less than at the bottom **(1)** and so its speed must be less as its mass has not changed. **(1)**

C Energy resources and electricity generation

Review questions (page 122)

1. Chemical energy (in fuel) is converted to heat energy by burning, used to provide kinetic energy (in high pressure steam, then in turbines then in generators), which is converted to electrical energy.

2. Sunlight can be focussed on a central furnace using mirrors, and the furnace heats water to make steam. Sunlight can be used to warm air beneath a transparent cover, which rises up a chimney and the moving air turns turbines which are used to drive generators.

3. (a) solar, wind, waves (as the sizes of the waves depend on the way the wind has been blowing)

(b) hydroelectricity, geothermal

(c) tides

Practice questions (page 122)

1. (a) The wind and sunlight will not be 'used up'/they will not run out. **(1)**

(b) (i) kinetic energy in wind → kinetic energy in generator → electrical energy **(1)**

(ii) chemical energy in diesel fuel → heat energy in engine → kinetic energy in engine → kinetic energy in generator → electrical energy **(1)**

(c) The wind does not always blow at speeds suitable for generating electricity. **(1)** Sunlight can only be used during the day **(1)** and when the sky is not too cloudy. **(1)**

(d) Burning the fuel adds carbon dioxide to the atmosphere **(1)** which is contributing to global warming/climate change. **(1)** Supplies of diesel will eventually run out. **(1)**

(e) (i) There may be times when tidal power is available when wind and solar energy are not available. **(1)**

   (ii) Tidal power is not available all the time. **(1)**

5 Solids, liquids and gases

A Density and pressure

Review questions (page 130)

1. $m = \rho \times V = 1200 \text{ kg/m}^3 \times 0.05 \text{ m}^3 = 60 \text{ kg}$

2. $A = \dfrac{F}{p} = \dfrac{60 \text{ N}}{50 \text{ Pa}} = 1.2 \text{ m}^2$

3. $\rho = \dfrac{p}{(h \times g)} = \dfrac{8000 \text{ Pa}}{(0.5 \text{ m} \times 10 \text{ N/kg})} = 1600 \text{ kg/m}^3$

Practice questions (page 130)

1. (a) $\text{volume} = \pi \times 0.25^2 \times 1$
   $= 0.196 \text{ m}^3$ **(1)**

   (b) $\text{density} = \dfrac{\text{mass}}{\text{volume}}$ **(1)**

   $= \dfrac{170 \text{ kg}}{0.196 \text{ m}^3}$ **(1)**

   $= 867 \text{ kg/m}^3$ **(1)**

   (c) $\text{pressure} = \text{height} \times \text{density} \times g$ **(1)**

   $= 1 \text{ m} \times 867 \text{ kg/m}^3 \times 10 \text{ N/kg}$

   $= 8670 \text{ Pa (or N/m}^2)$ **(1)**

   (d) $\text{area of bottom of barrel} = \pi \times 0.25^2$
   $= 0.196 \text{ m}^2$ **(1)**

   $\text{weight of oil and barrel} = 180 \text{ kg} \times 10 \text{ N/kg}$ **(1)**
   $= 1800 \text{ N}$ **(1)**

   $\text{pressure} = \dfrac{\text{force}}{\text{area}}$ **(1)**

   $= \dfrac{1800 \text{ N}}{0.196 \text{ m}^2}$

   $= 9184 \text{ Pa}$ **(1)**

B Change of state

Review questions (page 139)

1. (a) liquid

    (b) solid

    (c) liquid, gas

    (d) solid, liquid

    (e) gas

2. Evaporation can happen at any temperature. Boiling only happens at the boiling point of a liquid, and is evaporation happening as fast as is possible.

3. Your graph should look like this:

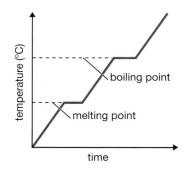

4. As the metal cools, the particles vibrate less and take up less space.

5. The temperature of the solid increases as it is heated until it reaches the melting point. The material is still being heated, but now the energy is being used to break bonds in the solid and allow a liquid to form. The temperature does not rise during the time that the solid is turning into a liquid. The energy needed to make this change is the latent heat of fusion. A similar thing happens when the material reaches its boiling point.

6. When sweat evaporates, it is the molecules in the sweat with the highest speeds that evaporate. The molecules left in the sweat therefore have a lower average speed, so the remaining sweat is cooler.

Practice questions (page 140)

1. (a) It flows/can be poured **(1)**, it takes the shape of a container. **(1)**

    (b) Each grain has a fixed shape **(1)** and volume. **(1)**

    (Being difficult to compress does not count for parts (a) and (b), as that property is common to solids and liquids.)

    (c) Close together **(1)** and vibrate about fixed positions. **(1)**

    (d) Close together **(1)** and move around randomly within the liquid. **(1)**

    (e) The fixed arrangement of particles is broken **(1)** so the particles can move around within the liquid. **(1)**

2. (a) Any two from the following points:

the particles in the material have more energy when they are hotter **(1)**
they vibrate more about their fixed positions **(1)**
sc the material takes up more space **(1)**

(b) The strip will bend downwards. **(1)**

(c) The strip acts as a switch which is closed in the position shown. **(1)**
When the strip warms up, it will bend away from the contact and open the switch **(1)** so the heating system will be switched off. **(1)**
The strip will straighten again when the room cools down **(1)** and switch the heating on again. **(1)**

(d) Reverse the metals **(1)** so that the switch closes the contact when it is warm. **(1)**
This will switch on the air conditioning when the room is too hot. **(1)**

3. (a) A higher temperature increases the rate of evaporation **(1)** because molecules in the liquid have more energy and so are more likely to be able to escape from the liquid. **(1)**

A greater surface area increases evaporation **(1)** as the molecules have more chance to escape from the liquid. **(1)**

A draught blowing over the liquid increases evaporation because it removes molecules in the air that have already evaporated **(1)** and makes it easier for more molecules to become a gas. **(1)**

(b) Water molecules that evaporate from the damp sand or pot have higher energies than the remaining water molecules **(1)** so the average energy of the remaining molecules is lower **(1)**
so the liquid water left is cooler. **(1)**
This cooler liquid can absorb heat energy from the food in the pot. **(1)**

C Ideal gas molecules

Review questions (page 146)

1. (a) 293 K

(b) −253 °C

2. $p_2 = \dfrac{p_1 T_2}{T_1} = 100\ 000\ \text{Pa} \times \dfrac{300\ \text{K}}{280\ \text{K}} = 107\ 143\ \text{Pa}$

3. $p_2 = \dfrac{p_1 V_1}{V_2} = 50\ 000\ \text{Pa} \times \dfrac{2\ \text{m}^3}{0.5\ \text{m}^3} = 200\ 000\ \text{Pa}$

Practice questions (page 146)

1. (a) Particles in a gas are moving all the time **(1)** and they exert a force on the walls of the container (or on something in the gas) when they bump into them. **(1)**

(b) If the gas is cooler the particles are moving more slowly **(1)** and so they hit the container walls less often **(1)** and with less force. **(1)**

(c) −273 °C **(1)**
At this temperature the particles stop moving and so there is no pressure. **(1)**
This temperature is called absolute zero. **(1)**

(d) 291 K **(1)**

2. 20 °C = 293 K **(1)**

$$T_2 = \frac{p_2 T_1}{p_1}$$

$$= \frac{2.5 \times 10^8 \text{ Pa} \times 293 \text{ K}}{5 \times 10^7 \text{ Pa}} \quad \textbf{(1)}$$

$$= 1465 \text{ K } \textbf{(1)}$$

Temperature at explosion = 1465 − 273

$$= 1192 \text{ °C } \textbf{(1)}$$

3. $V_1 = \dfrac{p_2 V_2}{p_1}$ **(1)**   (it is OK if you have all the 1s and 2s the other way round)

$$= 210 \text{ kPa} \times \frac{0.001 \text{ m}^3}{100 \text{ kPa}}$$ (you can use Pa or kPa, as long as all the pressures in the equation are in the same units) **(1)**

$$= 0.0021 \text{ m}^3 \textbf{(1)}$$

6 Magnetism and electromagnetism

A Magnetism

Review questions (page 151)

1. south to south, or north to north

2. Any three from: iron, steel, cobalt, nickel

3. Hard magnetic materials keep their magnetism after they have been magnetised, soft magnetic materials lose the magnetism quickly.

4. Stroke the pin repeatedly in the same direction.

5. Any two from: heat it, hit it repeatedly, put it into a coil of wire with alternating current flowing through it and turn the current down gradually.

6. (a) Your sketch should look like Fig. 6a.03.

    (b) Your sketch should look like the second part of Fig. 6a.04(a).

    (c) Your sketch should look like Fig. 6a.04(b).

    The key points for all three sketches are:

    - field lines curving around from one pole to the other
    - none of the lines crossing each other
    - arrows pointing from N to S.

Practice questions (page 152)

1. (a) The right-hand magnet should have N on it. **(1)**
        Your sketch should look like Fig. 6a.04(a).

    (1 mark for having the lines similar to Fig. 6a.04(a)
    1 mark for showing the arrows going away from the north poles of the magnets)

    (b) They would repel/move away from each other. **(1)**

    (c) The iron would be attracted/move towards the magnet. **(1)**

B Electromagnetism

Review questions (page 161)

1. Use a smaller current, use fewer turns of wire in the coil, remove a core made of magnetic material.

2. (a) Your sketch should look like Fig. 6b.01(a).

    (b) Your sketch should look like Fig. 6b.02(b).

    (c) Your sketch should look like Fig. 6b.04.

3. First finger is the magnetic field direction (from N to S).
    Second finger is the direction of the current (from + to –).
    Thumb is the direction of movement.

4. Increase the current, increase the strength of the magnetic field.

5. (a) Produce a uniform magnetic field.

    (b) Make electrical contact between the circuit providing the power and the moving coil.

    (c) Swap the connections over every half turn.

6. A coil of wire becomes an electromagnet when current flows through it, and attracts a piece of iron.
    When the iron moves, it closes the contacts in a separate circuit.

7. (a) Move the beam from side to side, to provide a time scale.

    (b) Move the beam up and down in response to a signal.

Practice questions (page 162)

1. (a) Your sketch should look like Fig. 6b.01(a).
        (1 mark for drawing circles around the wire, 1 mark for putting arrows on the circles)

    (b) Increase the current. **(1)**
        Increase the number of turns on the coil. **(1)**
        Use a core made of a magnetic material. **(1)**

    (c) When the switch is closed current flows **(1)** and the electromagnet attracts the armature. **(1)**
        This makes the striker hit the bell. **(1)**
        It also pulls the contacts apart so the current stops **(1)** and the armature springs back to its original
        position. **(1)**
        This completes the circuit again and the cycle repeats. **(1)**

    (d) soft **(1)**
        The electromagnet has to lose its magnetism quickly when the circuit is broken or the armature
        would not spring back quickly. **(1)**

2. (a) The disc spins anti-clockwise. **(1)**
   Conventional current flows from + to –, so down from the centre of the disc into the mercury. **(1)**
   The field goes from N to S, so into the page on this diagram. **(1)**
   Using the left hand rule, the motion caused is therefore to the right on the part of the disc between the ends of the magnet. **(1)**

   (b) Increase the current. **(1)**
   Increase the strength of the magnetic field. **(1)**

   (c) Make the current flow in the other direction. **(1)**
   Reverse the direction of the magnetic field. **(1)**

3. A: Heater, heats the cathode so it emits electrons. **(1)**
   B: Cathode, negatively charged to repel electrons. **(1)**
   C: Anode, positively charged to attract electrons and accelerate them. **(1)**
   D: Y-plates, given varying electric charges to deflect the electron beam vertically. **(1)**
   E: X-plates, given varying electric charges to deflect the electron beam horizontally. **(1)**
   F: Screen, glows where the electron beam hits it. **(1)**

C Electromagnetic induction

Review questions (page 169)

1. $V_p = \dfrac{V_s \times n_p}{n_s} = \dfrac{50\ V \times 50}{400} = 6.25\ V$

2. $V_s = \dfrac{V_p \times I_p}{I_s} = \dfrac{120\ V \times 20\ A}{0.5\ A} = 4800\ V$

Practice questions (page 170)

1. (a) zero **(1)**
   The magnet will stop moving when it is inside the coil, and a voltage/e.m.f./current is only induced when the magnetic field and coil are moving relative to one another. **(1)**

   (b) –2 A **(1)**
   The current will be in the opposite direction because the magnet is moving in the opposite direction. **(1)**

   (c) Increase the strength of the magnet. **(1)**
   Increase the speed of movement of the magnet. **(1)**
   Increase the number of turns of wire in the coil. **(1)**

   (d) In a generator the coil spins in a magnetic field **(1)**, and there are slip rings and brushes to connect the coil to the external circuit. **(1)**

2. (a) They change the voltage **(1)** of an alternating current. **(1)**

   (b) B, C, D, E **(1)**

   (c) $n_s = n_p \times \dfrac{V_s}{V_p}$ **(1)**

   $= 1000 \times \dfrac{400\ kV}{25\ kV}$ **(1)**

   $= 16\ 000$ **(1)**

(d) (i) Slightly less than 2 GW **(1)** as some energy will be wasted in the transformer. **(1)**
(Give yourself just 1 mark if you said the power in the 400 kV line is 2 GW)

(ii) The current is $\frac{1}{16}$ of the current from the power station **(2)**
(Give yourself 1 mark if you just said that the current is smaller)

The power is found by multiplying the voltage by the current **(1)** so if the power is approximately the same then if the voltage goes up the current must go down. **(1)**

7 Radioactivity and particles

A Radioactivity

Review questions (page 182)

1. $^{241}_{95}\text{Am} \rightarrow ^{237}_{93}\text{Np} + ^{4}_{2}\text{He}$

2. $^{137}_{55}\text{Cs} \rightarrow ^{137}_{56}\text{Ba} + ^{0}_{-1}\text{e}$

3. 90 minutes is 3 half-lives. The activity will be $120 \div 2 \div 2 \div 2 = 15$ Bq.

4. 120 Bq has to be halved twice to get 30 Bq ($120 \rightarrow 60 \rightarrow 30$), so 20 minutes represents 2 half-lives. The half-life is 10 minutes.

Practice questions (page 182)

1. (a) (i) atomic number/proton number **(1)**

(ii) number of protons in the atom **(1)**

(b) (i) mass number/nucleon number **(1)**

(ii) total number of protons and neutrons in an atom **(1)**

(c) isotopes **(1)**

(d) (i) around/outside the nucleus **(1)**

(ii) 8 **(1)**
An atom has the same number of protons and electrons. **(1)**

2. (a) (i) $^{4}_{2}\text{He}$

(ii) $^{0}_{-1}\text{e}$

(b) electrons **(1)**
ejected from the nucleus of atoms **(1)**

(c) $^{234}_{90}\text{Th} \rightarrow ^{234}_{91}\text{Pa} + ^{0}_{-1}\text{e}$

(1 for writing the symbol for thorium correctly
1 for the correct mass number for protactinium
1 for the correct atomic number for protactinium
1 for showing the beta particle correctly)

Answers

(d) $^{230}_{90}\text{Th} \rightarrow \, ^{226}_{88}\text{Ra} + \, ^{4}_{2}\text{He}$

(1 for writing the symbol for thorium correctly
1 for the correct mass number for radium
1 for the correct atomic number for radium
1 for showing the alpha particle correctly)

3. (a) Radon gas is a source of radioactivity **(1)** which can cause cancer or other health problems. **(1)**

    (b) Geiger–Müller detector **(1)**

    (c) Any two from: cosmic rays, internal radiation from within our bodies, food and drink, buildings and the ground, radioactive materials used in medicine, nuclear tests or accidents **(2)**

4. (a) The time it takes for half of the atoms in a sample to decay **OR** the time it takes for the activity of a sample to fall to half of its original value. **(1)**

    (b) Your graph should look something like this:

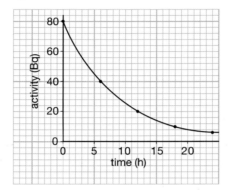

(1 for correctly labelled axes,
1 for plotting points at 80, 40, 20, 10 and 5 Bq,
1 for joining the points with a smooth curve)

    (c) It can be used to trace the movement of substances around the body. **(1)** A patient is injected with a substance containing a gamma emitter **(1)** and its progress around the body is followed using a camera sensitive to gamma rays. **(1)**

    (d) (i) Any one from: cancer, genetic mutations, birth defects **(1)**

        (ii) The film badge detects radiation **(1)** so it provides a measure of the amount of radiation the wearer has been exposed to. **(1)**

B Particles

Review questions (page 187)

1. (a) They expected the alpha particles to go straight through the gold foil.

    (b) The 'plum pudding' model of the atom said that the charges were spread out over the whole atom, so there was nothing with enough concentrated mass or charge to deflect an alpha particle.

2. (a) Some alpha particles passed straight through, but some were deflected by small angles, and some were deflected by large angles.

    (b) The alpha particles that went straight through passed through the gaps between nuclei. The ones that were deflected passed close to a nucleus, and the positive charges on the alpha particle and the nucleus repelled each other.

3. (a) It will be deflected more, as the same force on an object moving more slowly will cause a greater deflection.

   (b) There will be less charge on the nucleus so there will be a smaller force of repulsion and the deflection will be less.

4. (a) Slow down the neutrons so they cause more uranium nuclei to fission.

   (b) Absorb neutrons, to control how fast the chain reaction proceeds.

Practice questions (page 188)

1. (a) Most of the alpha particles passed straight through the foil **(1)** but some were deflected by small angles **(1)** and a few were deflected through large angles. **(1)**

   (b) Your sketch should look like Fig. 7a.01.

   (1 for drawing and labelling nucleus
   1 for drawing and labelling electrons
   1 for indicating that the nucleus has a positive charge)

   (c) The fact that most alpha particles passed straight through shows that most of an atom is empty space. **(1)** The large deflection of some particles shows that they were repelled by a small and massive part of the atom **(1)** that has a positive charge. **(1)**

2. (a) Two smaller nuclei **(1)** and some neutrons. **(1)**

   (b) Neutrons emitted when a uranium nucleus breaks up **(1)** hit other nuclei and cause them to break up. **(1)** These nuclei emit more neutrons which go on to break up other nuclei. **(1)**

   (c) Too many of the neutrons from a fission reaction escape from the material without causing further fission reactions. **(1)**

   (d) The fuel in the reactor is in the form of rods **(1)** surrounded by a moderator which slows down the neutrons to make them more likely to cause fission when they hit another fuel rod. **(1)** Control rods can be inserted into the moderator to absorb the neutrons **(1)** and the number of fusion reactions taking place can be reduced by inserting the control rods further into the reactor core. **(1)**

8 Practical work and investigations

Practice questions (page 206)

1. (a) Oil is thicker/more viscous than water **(1)** so the shapes would fall more slowly. **(1)**

   If the time is longer, any timing errors are a smaller proportion of the measurement (or any other way of explaining that using oil helps to reduce errors). **(1)**

   (b) The stopclock will be started as the shape passes the top line and stopped when it passes the bottom line. **(1)** Once a few objects have been dropped in, they will form a pile on the bottom of the tube, and so shapes tested later will not fall as far. Timing to the bottom line ensures that all shapes are timed over the same distance. **OR** It is easier to time from when the shape passes the top line than from when it is dropped (or similar explanation). **(1)**

(c) (i) 6.15 s for Shape A **(1)**

   (ii) The student may have forgotten to zero the stopclock after the previous measurement. **(1)**
   (Any other sensible suggestion would get the mark.)

(d) Mean for shape A = (3.12 + 2.87 + 3.05 + 2.91)/4 **(1)**

   = 2.99 s **(1 mark for correct value)**

Mean for shape B = (3.54 + 3.89 + 3.96 + 4.11 + 3.47)/5 **(1)**

   = 3.79 s **(1)**

(Plus 1 mark if both answers are given to 3 significant figures)

2. (a) Your graph should look something like this:

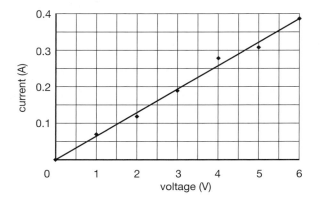

(1 mark for labelling the axes
1 mark for putting units on each axis
1 mark for plotting the points correctly
1 mark for drawing a line of best fit – not a line 'joining the dots')

(b) 0.1 A **(1)**

(c) $R = \dfrac{V}{I}$ **(1)**

   $= \dfrac{1.5\ V}{0.1\ A}$

   $= 15\ \Omega$ **(1)**

(d) Drawing a line of best fit is equivalent to finding a mean of repeated results **(1)** and allows any errors in individual readings to be cancelled out. **(1)**

(e) Any one of: same material of wire, same length of wire. **(1)**

(f) (i) It is difficult to tell if the curve is exactly the right shape to represent inverse proportionality. **(1)**

   (ii) Plot a graph of $\dfrac{1}{R}$ against area (or R against $\dfrac{1}{area}$). **(1)**

   If the relationship is one of inverse proportionality, the graph should be a straight line through the origin. **(1)**

3. (a) Any three from: the distance of the thermometers from the bulb, the amount of radiation emitted by the bulb (by keeping the same voltage setting on the power supply), the temperature of the air in the room, any draughts or air currents in the room. **(3)**

(b) With Group B's method, the temperature or draughts in the room could be different for the two thermometers. **(1)**
    With Group A's method it is easier to ensure that the two thermometers are exactly the same distance from the bulb. **(1)**

(c) They have joined all the data points instead of drawing lines of best fit. **(1)**

(d) (i) The temperature of the black-painted thermometer increased at a greater rate/faster than that of the white-painted thermometer **(1)** showing that black surfaces absorb radiation faster/better than white surfaces. **(1)**

   (ii) The temperature of the black-painted thermometer decreased at a greater rate/faster than that of the white-painted thermometer **(1)** showing that black surfaces emit radiation faster/better than white surfaces. **(1)**

(e) Any anomalous results will be obvious because they will lie off the line of the other results on the graph (or similar explanation). **(1)**

(f) t is clear from the graph that the temperature of the black-painted thermometer is decreasing faster than that of the white-painted one. **OR**
    The temperature of the black-painted thermometer continues to decrease faster than that of the white one even after it has cooled down to the maximum temperature reached by the white one. **(1)**

Appendices

1

Formulae

There are many different formulae that you need to be able to use for International GCSE Physics. Using a formula means being able to substitute the correct numbers into it, but it also means being able to rearrange the formula to change its subject.

Some of the formulae you need to be able to use will be provided in the exam paper, but many will not. The table at the end of this appendix (on page 241) lists all the formulae that you will need to memorise.

Formula triangles

If you are confident at algebra, you should have no difficulty rearranging the formulae that you may need to use in the examination. If you are not confident, you can memorise a formula triangle for most of the equations. If you do this, it is *very important* that you remember the correct places for all the letters in the triangles.

The triangle below shows the formula for calculating speed. To use it, cover up the quantity you need to calculate, and the remaining letters give you the formula:

- to calculate speed, cover up the s and what you can see is $\dfrac{d}{t}$ so speed $= \dfrac{\text{distance}}{\text{time}}$

- to calculate distance, cover up the d, and what you can see is $s \times t$, so distance = speed × time.

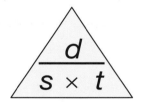

Rearranging formulae

This section gives you some reminders about how to rearrange formulae. It does not explain *why* these methods work – you should have learnt this in maths lessons.

Easy rearrangements

The easiest formulae to deal with have three terms in them, like the speed formula. The trick with these formulae is to move things 'top to bottom'.

To calculate *distance*, move *time* from the bottom of one side of the formula to the top of the other side:

$$\text{speed} = \frac{\text{distance}}{\text{time}}$$

$$\text{time} \times \text{speed} = \text{distance}$$

This is just the same as: distance = speed × time

If you want to calculate *time*, swap *speed* with *time*:

$$\text{speed} = \frac{\text{distance}}{\text{time}}$$

$$\text{time} = \frac{\text{distance}}{\text{speed}}$$

Check these with the formula triangle opposite.

The same method works if there is no division in the original formula:

work = force × distance moved in direction of force

or $W = F \times d$

To calculate *F*:

$$W = F \times d$$

$$\frac{W}{d} = F$$

To calculate *d*:

$$W = F \times d$$

$$\frac{W}{F} = d$$

The method still works when there are three terms on one side of the formula, such as the formula for calculating gravitational potential energy:

$$\text{GPE} = m \times g \times h$$

To calculate *m*:

$$\text{GPE} = m \times g \times h$$

$$m = \frac{\text{GPE}}{g \times h}$$

Slightly harder rearrangements

The transformer equation can be rearranged using the same rules. It just looks a little trickier.

$$\frac{V_p}{V_s} = \frac{n_p}{n_s}$$

Finding V_p or n_p is quite easy:

$$\frac{V_p}{V_s} = \frac{n_p}{n_s}$$

$$V_p = \frac{n_p}{n_s} \times V_s$$

$$\frac{V_p}{V_s} = \frac{n_p}{n_s}$$

$$n_s \times \frac{V_p}{V_s} = n_p$$

which is the same as
$$n_p = \frac{V_p}{V_s} \times n_s$$

To find V_s or n_s you need to do two swaps ...

$$\frac{V_p}{V_s} = \frac{n_p}{n_s}$$

$$\frac{n_s \times V_p}{n_p} = V_s$$

... or just turn the formula upside down and do a single move:

$$\frac{V_s}{V_p} = \frac{n_s}{n_p}$$

$$V_s = \frac{n_s}{n_p} \times V_p$$

Harder still

$$KE = \frac{1}{2} \times m \times v^2$$

You can find m using the method above.

To find v, first rearrange the formula to find v^2, then take the square root:

$$KE = \frac{1}{2} \times m \times v^2$$

$$\frac{KE}{\frac{1}{2} \times m} = v^2$$

$$v = \sqrt{\frac{KE}{\frac{1}{2} \times m}}$$

Learning the formulae

The table below shows all the formulae that you need to learn for the examination. These formulae will *not* be given to you on the exam papers.

In words	In symbols	Triangle	Page
average speed = $\dfrac{\text{distance}}{\text{time}}$	$s = \dfrac{d}{t}$	$\dfrac{d}{s \times t}$	3
force = mass × acceleration	$F = m \times a$	$\dfrac{F}{m \times a}$	16
acceleration = $\dfrac{\text{change in velocity}}{\text{time taken}}$	$a = \dfrac{(v - u)}{t}$		6
density = $\dfrac{\text{mass}}{\text{volume}}$	$\rho = \dfrac{m}{V}$	$\dfrac{m}{\rho \times V}$	125
work done = force × distance moved in direction of force	$W = F \times d$	$\dfrac{W}{F \times d}$	110
kinetic energy = $\frac{1}{2}$ × mass × speed²	$KE = \frac{1}{2} \times m \times v^2$		112
gravitational potential energy = mass × g × height	$GPE = m \times g \times h$	$\dfrac{GPE}{m \times g \times h}$	111
weight = mass × gravitational field strength	$W = m \times g$	$\dfrac{W}{m \times g}$	17
pressure = $\dfrac{\text{force}}{\text{area}}$	$p = \dfrac{F}{A}$	$\dfrac{F}{p \times A}$	127
moment = force × perpendicular distance from pivot	$M = F \times d$	$\dfrac{M}{F \times d}$	28
charge = current × time	$Q = I \times t$	$\dfrac{Q}{I \times t}$	62
voltage = current × resistance	$V = I \times R$	$\dfrac{V}{I \times R}$	61
electrical power = voltage × current	$P = V \times I$	$\dfrac{P}{I \times V}$	46
wave speed = frequency × wavelength	$v = f \times \lambda$	$\dfrac{v}{f \times \lambda}$	76
$\dfrac{\text{input (primary) voltage}}{\text{output (secondary) voltage}} = \dfrac{\text{primary turns}}{\text{secondary turns}}$	$\dfrac{V_p}{V_s} = \dfrac{n_p}{n_s}$		167
refractive index = $\dfrac{\sin(\text{angle of incidence})}{\sin(\text{angle of refraction})}$	$n = \dfrac{\sin i}{\sin r}$	$\dfrac{\sin i}{n \times \sin r}$	87
sin(critical angle) = $\dfrac{1}{\text{refractive index}}$	$\sin c = \dfrac{1}{n}$		91
pressure difference = height × density × g	$p = h \times \rho \times g$	$\dfrac{p}{h \times \rho \times g}$	128

2

Electrical circuit symbols

The table below shows the electrical circuit symbols that you will be expected to recognise and use. Some of these are mentioned in the main chapters of this revision guide, because you need to understand how they work or how they are used.

Description	Symbol	Page
conductors crossing with no connection		64
junction of conductors		51
open switch		51
closed switch		158
open push switch		
closed push switch		
cell		51
battery of cells		156
power supply	(d.c.) or (a.c.)	60
transformer		166
ammeter	A	56
milliammeter	mA	
voltmeter	V	54
fixed resistor		52
variable resistor		60

Description	Symbol	Page
heater		60
thermistor		57
light-dependent resistor (LDR)		57
relay		158
diode		55
light-emitting diode (LED)		55
lamp		51
loudspeaker		158
microphone		
electric bell		162
earth or ground		159
motor	M	157
generator	G	165
fuse/circuit breaker		44